The Hacker Diaries:

Confessions of Teenage Hackers

Dan Verton

McGraw-Hill/Osborne

New York   Chicago   San Francisco
Lisbon   London   Madrid   Mexico City   Milan
New Delhi   San Juan   Seoul   Singapore   Sydney   Toronto

McGraw-Hill/Osborne
2600 Tenth Street
Berkeley, California 94710
U.S.A.

To arrange bulk purchase discounts for sales promotions, premiums, or fund-raisers, please contact **McGraw-Hill/**Osborne at the above address. For information on translations or book distributors outside the U.S.A., please see the International Contact Information page immediately following the index of this book.

The Hacker Diaries: Confessions of Teenage Hackers

1234567890 DOC DOC 0198765432

ISBN 0-07-222364-2

1234567890 DOC DOC 0198765432

ISBN 0-07-222552-1

Publisher	Brandon A. Nordin
Vice President & Associate Publisher	Scott Rogers
Acquisitions Editor	Jane K. Brownlow
Project Editor	LeeAnn Pickrell
Acquisitions Coordinator	Emma Acker
Copy Editor	Judy Ziajka
Proofreader	LeeAnn Pickrell
Indexer	Claire Splan
Illustrators	Michael Mueller, Lyssa Wald
Series Design	Jean Butterfield
Cover Design	Jeff Weeks

This book was composed with Corel VENTURA ™ Publisher.

To Mom, Dad, and Corinne
Life, strength, and love

About the Author

Dan Verton is a senior writer and veteran investigative reporter with *Computerworld* in Washington, D.C.

As one of the leading technology journalists in the country, Verton traveled around the world in 1999 and 2000 covering the NATO-led war in Kosovo and the use of cyberwar tactics by the U.S. military. He has interviewed and written profiles of hackers, Kosovo refugees, and military cyber-warriors from around the world, and has traveled with the U.S. Secretary of Defense. His career has taken him from the front lines of military cyber operations to the front lines of hacker wars in the U.S.

Verton is a former intelligence officer in the U.S. Marine Corps. He earned an M.A. in Journalism from American University in Washington, D.C., a B.A. in History from the State University of New York-Binghamton, and has attended the University of Pennsylvania.

CONTENTS

Acknowledgments   vii
Introduction   ix

1 "Genocide": From Columbine to Hacking   1

2 Rebels: Joe Magee and "Noid"   27

3 The Hunt for Mafiaboy: Operation Claymore   55

4 A Tale of Two Script Kiddies: Pr0metheus
and Explotion   90

5 World of Hell 109

6 Cyberchic: Starla Pureheart 127

7 Unlikely White Hat: Willie Gonzalez 142

8 Tinker, Teenager, Hacker, Spy:
The H.D. Moore Story 166

Afterword 182

A Two Decades of Teenage Hacking 195

B Making Headlines Over the Years 203

C Hacking on the Web 207

Index 213

ACKNOWLEDGMENTS

The McGraw-Hill Osborne/team who worked on this book includes Jane Brownlow, senior acquisitions editor, Emma Acker, acquisitions coordinator, Judy Ziajka, copy editor, LeeAnn Pickrell, senior project editor, Lyssa Wald, illustration supervisor, and Bettina Faltermeier, publicity manager. There were undoubtedly countless others throughout the editorial and production departments at McGraw-Hill/Osborne who played a significant role in making this project happen. All did a wonderful job and improved the quality of the book immensely.

I also want to acknowledge all of the hacking and security professionals who provided me with their time and expertise. In particular, I want to thank Raphael Protti of The Loomis Group, Thubten Cumerford of White Hat Technologies, Inc., Michael Turner of McAfee, and Phyllis Schneck of SecureWorks, Inc.

Robert R. Butterworth, Ph.D., of International Trauma Associates, also provided critical insights into the psychology of today's teenage mind. Likewise, Christopher Alghini, a former technology director and computer science teacher for Berkshire School in Sheffield, Massachusetts, offered wise counsel.

A special thanks goes to Brian Martin (a.k.a. "Jericho"), one of the founders of Attrition.org, for providing first-hand knowledge of the hacking and defacing scene and offering the full use of the Attrition.org defacement mirror. Likewise, thanks goes to the staff of the Allas online defacement mirror, for allowing me to use defacements from their site as well.

Lastly, I want to thank all of the hackers, security officials, and law enforcement officers from the FBI and the Royal Canadian Mounted Police who took countless hours out of their lives to talk with me and exchange what must have seemed like an endless stream of e-mails. All expressed their support for such a book and each dove into this project with as much enthusiasm as I did. I am forever grateful for their willingness to share their lives with me.

Nothing will ever be attempted,
if all possible objections
must first be
overcome.

Integrity without knowledge is weak and useless,
and knowledge without integrity is
dangerous and
dreadful.

Be not too hasty ... to trust, or to admire, the teachers of morality;
they discourse like angels, but they live like men.

—*Samuel Johnson, "Rasselas," 1759*

INTRODUCTION

Wednesday, March 25, 1998
One Month After My Arrest

Dear Diary:

Don't believe what you read in the newspapers. This is how it really happened.

People think they know me, but they don't. They know only what the government and the media want them to know. I'm not a terrorist or a murderer. Hell, I'm not even 19 years old. But I'm pretty good with a computer. Actually, I'm very good, better than most people. I'm what they call a hacker. Don't ask me if I'm a "black hat" or a "white hat" hacker—labels are for lamers, those who don't know what hacking is really all about. Black or white, bad or good. Who cares? No label can define who I am, what I've done, or why I did it. Not even my lawyer can figure that out. But I will tell you that I'm misunderstood and stand wrongly accused of crimes that are not crimes at all.

The charges levied against me are bogus. I remember how ridiculous it all sounded when the judge read them aloud: "unauthorized access to a computer system, felony vandalism, and conspiracy to commit telecommunications fraud and computer hacking." There were others, but I stopped listening after those first few. If you want the truth, my crime is my addiction to knowledge, love of technology, and belief in the inherent freedom of information. I'm a problem solver who takes pride in creating problems for the ignorant. That's the only way the ignorant learn. I'm an Internet-age Marco Polo in search of new adventures. I'm a malcontent who's fed up with the system, especially school, where the computer teachers can't teach me what they don't know. I'm also curious, shy, popular, athletic, lazy, rich, poor, and friendly. I walk dogs after school for the elderly people in my neighborhood, and on Saturdays I bag groceries at the local supermarket. I look like every other teenager around here. I listen to the same music and attend the same schools. I get the same grades as all of the other kids, and yet sometimes I barely pass my classes. And you, you who believe everything you read about hackers, think you know me. You don't know me.

I conducted my first serious hack last summer. That was what started me down the road that led to where I am now, which is a world of legal trouble. I won't get into all of that lawyer mumbo-jumbo right now, but I will tell you that this time there's no getting out of it—trouble that is. I admit I'm a little scared.

Anyway, I had been hacking and teaching myself how to break into computers for about four years before I got involved in the Web page defacing scene last summer and started to do more serious things. I'll get to that shortly. I started like every hacker starts: I taught myself how to program and searched the Internet for information on how to make my computer do what I wanted it to do. I was about 14 at the time. Searching the Internet is like walking through the world's largest library, only there are no separate sections for adults and minors. Everybody has equal access. On the Internet, if you're smart enough to find the information, it's yours.

You name it and I found it on the Internet. There were more hacking texts available than I could read in a lifetime, but I read them all anyway. There are no shortcuts to becoming a hacker. I'll never forget the first anarchy text I ever read. It was called the Anarchist's Cookbook. *I didn't know at the time that most of the information was out of date, but I thought it was really cool to have a book that taught you not only how to hack, but also how to pick locks, make bombs, produce counterfeit money, make letter bombs, steal calls from pay phones, and my favorites, how to kill someone and send a car to HELL.*

I didn't learn much about hacking from the Anarchist's Cookbook, but just having that type of information in my possession gave me the feeling, for the first time in my life really, that I belonged to something larger than myself. I had become a member of the "electronic underground" that I had heard so much about. The underground. That's a funny term, isn't it? Do you know that there are people everywhere who read news stories about hackers and actually think that we live under the ground? They think we dwell in the back alleys and sewers of towns far away, like the dirty, sweaty villains they see on television. Most people have no idea that members of the so-called hacker underground sit one pew away from them in church every Sunday, process their credit cards at their favorite stores in the mall, play basketball on the same team as their sons or daughters, and work in sensitive areas of businesses all over the country, even businesses that specialize in, of all things, computer security. Scary, huh? Yeah, but it's true.

My hacking education shifted into high gear when I discovered this thing called IRC. That stands for Internet Relay Chat. Hackers and Internet surfers use IRC to talk to each other. It's like using a telephone only with letters on a computer screen. It was on IRC that I first met DarkViper and Prophet. We met in this special chat room where only hackers hang out. I used the hacker nickname Skeleton. It was a nickname that one of my friends at school gave me as a joke. I was always a skinny kid, so some of my friends called me Skeleton, or Skel, or Mr. Skeleton. We had names for everybody. You shouldn't get hung up on the names we hackers pick for ourselves. Very few of them really mean anything. They're chosen mostly for their shock value or because they just sound cool. Don't get all psycho on me and try to analyze my personality based on my hacker nickname. If you really want to hear a bunch of sick nicknames, just go to a professional boxing match and listen as the announcer introduces "The Executioner" or the "Bayonne Bleeder." You also could go to a professional wrestling match and watch "Suicide Blonde" take on "Sex and Violence." Nobody thinks twice about the nicknames those guys pick.

But I digress.

DarkViper and Prophet were regulars in the IRC scene. At first, they responded to my questions about hacking by telling me to go learn on my own like everybody else did. Eventually, I was able to prove that I knew what I was talking about and started to share some hacking files with them. Soon, the three of us became like real friends. That was a little strange because we had never even met each other in person or talked to each other on the telephone. IRC at two o'clock in the morning was our common bond.

Prophet became my mentor. We exchanged code and helped each other refine our techniques. He taught me how to use various commands and how to mask my computer's Internet Protocol (IP) address on the Internet. IP addresses are like street addresses for computers, and not learning how to hide your true identity is a good way to get caught. So I learned how to do that early in my career. Prophet also taught me how to find password files on servers.

DarkViper supplied me with my hacking scripts and software tools and was my primary teacher when it came to operating systems and learning how to code in different programming languages, such as C and Perl. He gave me a few programs that are designed to crack passwords and helped me learn how to use them. This was a welcome development because I couldn't afford to get banned from another Internet service provider (ISP). I had tried, unsuccessfully, to dupe stupid, gullible people into e-mailing me their passwords by impersonating ISP technicians. But one more phone call from the ISP, and my parents would have cut me off for good.

I don't know what to say about my parents. Before I got arrested, they had no clue about what I was doing online. And it's not like they can't use a computer. My father's a technician for an airline, and my mother (she's actually my stepmother) is an accountant. Both of them know how to use a computer. But neither of them really cared too much about what I was doing. In their minds, it was better that I was at home on the computer than out doing drugs or something illegal. Our relationship has always been sort of loose; no curfew, no rules about computer use, nothing. And I'm not overly needy. I don't ask for too much, so they usually give me whatever I want. I always want new computer equipment.

I also was good at concealing my hacking from them. I don't know any hackers my age who aren't good at keeping their parents in the dark. I always had a video game running in the background or a sports news Web page waiting just in case one of them knocked on my bedroom door out of the blue. Sure, they were a little suspicious about the ISP's banning my account, but I was pretty much able to convince them that it was all a big mistake.

High school was always a good place to refine my hacking skills. But don't misunderstand what I'm saying. School was a good place to hack not because the teachers knew a lot about coding or networking or security, but because you could hack somebody else's account, and bingo—you had plausible deniability that you weren't the person sitting at the computer when the hack took place. At my school, the teachers were so lame and so far behind the knowledge power curve that I often had to hack just to challenge myself. There was a reason I always sat in the back of the computer class.

By the time I was 17, I was constantly scanning for open ports to computers on the Internet. My hacking practice, as I liked to call it, lasted anywhere from 6 to 8 hours a night on the weekdays and up to 12 hours on the weekends. At first, I shied away from hacking into systems that belonged to big companies and the government. Instead, I concentrated on obscure Web pages belonging to companies that nobody had ever really heard of. I could gain root access to these servers (which means I was basically God on the system) so easily that I usually just broke the connection and moved on. Don't get me wrong; I could have destroyed everything on the system, rummaged through credit card numbers, whatever. But they were so easy to crack that it took all of the fun and all of the rush out of it. It felt like I was mugging an old lady.

That's when the three of us—me, DarkViper, and Prophet—decided to form a group. Since it was my idea to form the group, we called ourselves The Skeleton Crew. I thought that was a cool name because it also was the name of an awesome collection of short stories by Stephen King (Imagine that! A hacker who reads books.) We all were fed up with the complete lack of real security on the Internet. People always fail to realize that hackers are people, too. We hate it as much as the next person when our credit cards are stolen because some administrator at an Internet company didn't know how to protect the system. The Skeleton Crew's mission, therefore, was to expose the ignorant for what they were. Then maybe people would start to take security seriously. And then maybe one day they would hire us to do it for them.

The Skeleton Crew started simple. We scanned the Internet for systems that had well-known vulnerabilities and that we had already written scripts to take advantage of. It usually took us a few minutes to get in. The script would exploit the system at the click of a mouse and drop us in as root. Then we would simply replace the Web site's home page with a message that told the system administrator that their server was wide open to attack. We never destroyed any information. In fact, we almost always saved the original Web page and told the administrator where they could find it. We did this for a while and hit dozens of Web sites. I always loved it when I came across a site that was a little tougher to hack than most. Sometimes I would work on a site for weeks trying to figure out how to get in. I always got in. Nothing is 100 percent secure. Not even from me, a teenage "script kiddie."

The Skeleton Crew was now on the map. We felt like we were the kings of the Web page defacing scene. We didn't just hit companies that used Microsoft Windows systems. That would be too easy. We were pros. The Skeleton Crew could hit any operating system: Unix, Windows, Linux, Irix, whatever. We approached everything *we did seriously, even though we did it mostly for fun and for the rush of being able to defeat systems that others said were secure.*

It wasn't long, however, before we got bored with the whole scene. Every script kiddie who could hack a Windows system thought he was a bad ass. But their mouths were always bigger than their hacks. So we in The Skeleton Crew decided it was time to separate ourselves from the mindless horde of wannabes. And that's when all of the trouble I mentioned earlier started.

Prophet came up with the idea to go on a rampage through the networks at the Pentagon. It just so happens that this was at the same time that the Pentagon was planning a new military campaign against Iraq. Prophet said he had already mapped out a plan that would enable us to make it look like the hacks were originating from points all over the world. We would be able to keep them guessing and off balance enough to work our magic quietly and stealthily on other systems. Then we could use those systems to come and go as we pleased, and nobody would ever know. But in the end, The Skeleton Crew would be able to rack up the most military Web site defacements of all time and show the world what the best were really capable of. I thought this was cool and knew right away that it would take DarkViper and me to a new level in our hacking.

We spent about a week gaining access to systems all over the world, including Israel, the United Arab Emirates (UAE), France, Taiwan, Mexico, and Germany. We used these systems to launch the attacks against the Pentagon. First we scanned for systems that had a well-known vulnerability in the Remote Procedure Call (RPC) code of the Solaris operating system. This allowed us to execute commands and programs on the target systems remotely. We also installed sniffer programs that enabled us to capture hundreds of network passwords. The passwords helped us gain root and system administrator level access to at least 500 military domain name servers—the major systems that route data requests between computers on a network.

Armed with the all-powerful root and system administrator accounts, DarkViper and I installed trap doors, which positioned us to be able to crash the systems and send the Pentagon's military operations into a tailspin. We didn't do anything stupid, of course. But we were a little surprised that we hadn't received any press coverage yet. It was like nothing was happening.

We didn't know it at the time, but behind the scenes the Pentagon had formed a task force of experts from the Air Force, the FBI, the National Security Agency, the CIA, and the Justice Department to investigate what we were doing. I didn't find out until much later that representatives from the task force briefed the Deputy Secretary of Defense daily on the situation. Congress even held a classified hearing on the hacks. They thought we were Iraqi intelligence agents. Heh!

We really had them guessing. Prophet was the brain behind most of the operation. He walked DarkViper and me through almost every stage of the attacks. With his help, we penetrated systems at seven Air Force bases and four Navy installations, the Department of Energy's National Laboratories, where nuclear weapons research is conducted, NASA sites, and various university networks. On average, we hopped between eight different systems before attacking our ultimate Pentagon targets. We did this for a month.

I'll never forget the day it all came crashing down around us. There must have been two-dozen FBI agents banging on the front door to my house that morning. All I remember hearing was the sound of a deep voice yelling "search warrant" and my stepmother's racing heart pounding in my own ears, gradually drowning out all of the commotion.

The agents walked right into my bedroom as if they knew exactly where they were going. They confiscated everything: both of my computers, printers, and every floppy disk and CD they could find. Oh yeah, and they also confiscated me.

I later learned that the investigation into our hacking spree was given a code name. The feds called it Operation Solar Sunrise. As it turns out, the FBI is more interested in getting their hands on Prophet, who, they say, is one of the best hackers in the world. I don't doubt that, but it was a surprise to learn that he actually lives in Israel. I guess I didn't know him as well as I thought.

As for DarkViper and me, we're screwed. Our futures are in the hands of a judge. Everything I was working for is in jeopardy. I have no hope that this judge will understand why I did what I did, and even less hope that he'll understand what hackers are all about. I'm not even sure that I understand anymore.

To be continued…

^ ( ) < [ ] > ( ) *^*

The characters in the diary entry you just read are fictional. However, all of the events it describes actually happened or are based on my personal interviews with real teenage hackers. It reflects a story that has been repeated countless times during the past 10 years. And it describes teenagers who live and breathe in every neighborhood in the country, including yours.

Solar Sunrise was neither the first nor the last operation in which teenage hackers would become the focus of a massive manhunt by the FBI. In 1983, the FBI arrested seven Milwaukee-based computer hackers known as the 414s in one of the first major hacker stings to gain national notoriety. The 414s, named after their local telephone area code, were convicted of breaking into more than 60 systems, including Security Pacific National Bank and the Los Alamos National Laboratory. All seven of the 414s were teenagers.

It happened again in 1990, when the FBI launched a 12-state hacker crackdown known as Operation Sundevil. The arrests came on the heels of the 1990 crash of the AT&T long-distance telephone network attributed to a group of teenage hackers who called themselves the Masters of Deception. The MOD boys, three teenagers from Queens, New York, were indicted in 1992.

The most recent international manhunt to be launched in search of a teenage hacker occurred in February 2000, when the FBI and Royal Canadian Mounted Police arrested a 15-year-old Montreal boy who was responsible for bringing down some of the biggest e-commerce companies on the Internet. At the time of this writing, a similar manhunt is underway for the author of the Code Red worm, which infected hundreds of thousands of computers around the world and threatened to slow Internet communications to a crawl.

But these are only some of the stories, the ones that have made the headlines. They don't come close to telling the behind-the-scenes life stories of all teenage hackers. Although good hackers start young, they cannot be profiled. There is no one picture of the average teenage hacker. They are the kids bagging your groceries at the supermarket; working in community service on the weekends; mowing their parent's lawns; competing on the high school wrestling team; playing an instrument in the school orchestra or singing in the choir; taking art lessons and reading military history; struggling with their grades in math, science, and English; getting good grades and planning a bright future; hanging out with their friends after school and getting into trouble, sometimes even getting arrested; and almost always feeding their obsession with computers and the Internet late at night.

This book tells the life stories of the rank and file of the hacker subculture, the teenage hackers who truly define the scene. More than just a series of stories about the technical aspects of their hacking and cracking exploits, this book explores why these average teenagers got involved in hacking in the first place, how they think, what life was like for them growing up, the internal and external pressures that pushed them deeper and deeper into the hacker underground, and what they found once they got there. It also is a book that for the first time breaks down the stereotypes that have shaped the public's perception of the teenage computer hacker as a dangerous, murderous villain. Very few teenage hackers are the social misfits with pimpled faces that the media makes them out to be and that the general public fears as much as it fears society's hardened criminals. To the contrary, many are athletic, well liked, indistinguishable from the rest of their friends, and increasingly female.

The real teenage hacker culture is a patchwork of different personalities, backgrounds, motivations, experiences, and opinions. For example, there is Genocide. The kid with the controversial hacker nickname grew up in a small wooden shack in Alaska that had no electricity, telephone, or running water. Still, as we see in Chapter 1, the poor kid from Alaska managed to form a hacker group that would end up on the FBI's radar screen.

And there are the curious renegades represented by Joe Magee and Noid. Although as teenagers they were polar opposites in terms of personality, Joe and Noid shared an insatiable curiosity about technology, each using his family's VCR to launch a teenage career in hacking and cracking computers. We meet them in Chapter 2.

Then there's Mafiaboy, perhaps the most notorious teenage hacker of the last few years. The product of a dysfunctional family environment, Mafiaboy single-handedly caused more than a billion dollars in damage to some of the biggest companies on the Internet. Only 15 when he launched a devastating series of denial-of-service attacks in February 2000, Mafiaboy demonstrated clear criminal intent. He was a 125-pound brash loudmouth and ever since his arrest has shown little or no remorse for his actions. He's a big part of the reason people have come to fear teenage hackers. And, ironically, he wasn't a very good hacker. Chapter 3 offers a behind-the-scenes look at Operation Claymore, the joint operation conducted by the FBI and the Royal Canadian Mounted Police to track down and prosecute Mafiaboy.

In Chapter 4, we explore the lives of two classic script kiddies, Pr0metheus and Explotion. Both are angry, but for different reasons, and not very good reasons. Pr0metheus is a disciple of Satan, an active hacker who defaces Christian and religious Web sites as part of an angry, personal war against organized religion. Explotion's personal war is internal. The 19-year-old has yet to decide where he fits in the larger scheme of white-hat, gray-hat, and black-hat hackers. But this much he knows for sure: most people are idiots who go out of their way to annoy him.

Most script kiddies do not act alone, however. In Chapter 5, we get up close and personal with World of Hell, one of the most notorious Web site defacement gangs in recent history, and we learn how they banded together. In addition, we get a rare glimpse into how talented and at times tormented many of these kids are.

There's also some light at the end of the tunnel. Take, for example, 15-year-old Anna Moore (a.k.a. Starla Pureheart), the first female hacker to win the ethical hacking contest at the annual DefCon hacker conference in Las Vegas. Profiled in Chapter 6, Anna is more than a symbol and a leader for the growing number of women joining the underground; she's a shining example of sophistication and ethical behavior. She's no Mafiaboy; she's got real "skilz" and a moral compass that is enforced by what some teenager hackers might consider an overbearing mother. Regardless, Anna has definite opinions of her own about the hacking scene of which she is a part.

The same can be said about Willie Gonzalez, a teen hacker who had the capabilities to do great damage, but who, like Anna Moore, also had the moral courage to walk away from the dark side before it was too late. In Chapter 7, we look at how Willie's early life as a lonely outsider who turned to computers and hacking as an escape helped form his character as a young adult. Likewise, an older Willie Gonzalez found himself in the unique position of mentoring another young teenage hacker, a responsibility he did not take lightly.

Although all of the hackers interviewed for this book are truly unique, none tells a more interesting and compelling story than H.D. Moore, profiled in Chapter 8. Moore, who started hacking at the age of 13 as an escape from being shuffled between divorced parents, kicked out of schools, and running with members of some of the toughest street gangs in Austin, Texas, is considered by some who know him to be one of the best hackers in the world. By the age of 16, H.D. had received promises of future job references from major companies for having discovered serious vulnerabilities in their Web sites. By 17, he had worked with a special Navy security research group, authored well-known tools for analyzing hacker attacks, given a live presentation to a group of thousands of system administrators and security professionals as a recognized hacking expert, and accepted a job working in support of classified Air Force information warfare programs. Now, at 20, H.D. is a respected member of the vulnerability assessment team at Digital Defense, Inc., in San Antonio, Texas.

Are teenage hackers damned to a life on the fringe, a life of crime? Are their reputations forever tarnished by their youthful electronic transgressions? And what about those who get involved in more serious hacking? What about those who knowingly break the law? Are they selling out their futures, or, in the case of Pr0metheus, are they selling their souls? Will companies hire them to work as computer security gurus? Will anybody hire them? Have any former teenage hackers made the transition from hacker to security consultant successfully?

Seldom do we get a look at how the mainstream teenage hacker community lives, how they grow up, and what they think about the notorious subculture of which they are a part. The hacker underground cannot be defined by headlines in a newspaper or magazine or endless ranting on Internet message boards. It's too complicated for that. It's real life, and real people are living it.

A word is in order about how the lives of the hackers you are about to meet were reconstructed. All agreed that a book like this was both necessary and long overdue. As a result, they endured what must have seemed to them like endless conversations and e-mail exchanges. However, many agreed to share their personal life stories as hackers on the grounds that I not disclose their real identities. I have honored those requests, referring to some only by their hacker screen names and in one other case changing the individual's name completely.

This is a story of technological wizardry, creativity, and dedication; of youthful angst, boredom, and frustration; of disconnection from society, anger, and jail time. It is the story of today's teenage computer hackers. They're not the monsters that we read about. They're just like every other kid, and some of them probably live in your neighborhood. They're there. All you have to do is look.

Dan Verton
Washington, D.C.
2002

1

"Genocide": From Columbine to Hacking

In high *school, they were known by their nicknames, "Reb" and "VoDKa." They were quiet, shy and gentle, and content to be by themselves. They played video games, collected baseball cards, and watched movies. They worked together producing videos for the school's news network. As young boys, they served as Boy Scouts and played Little League baseball and soccer.*

They were smart too; VoDKa took part in the high school's gifted student program and, along with Reb, had acquired a keen interest in computers. VoDKa even built his own computer at home and dreamed of becoming a computer science major in college. Both boys worked as system administrators for the school's computer lab and maintained their own Web pages.

But 18-year-old Eric Harris and 17-year-old Dylan Klebold were definitely not like other teenagers. They were outsiders who were filled with violence, hatred, and a blind devotion to Adolf Hitler's Nazi rhetoric. On April 20, 1999, that blind hatred erupted, not with an online rampage through the school's computer network, but with double-barrel shotguns, semiautomatic rifles and pistols, 30 homemade hand grenades, pipe bombs filled with nails, and a propane tank rigged with explosives. The four and a half hours of chaos that Harris and Klebold unleashed on that day at Columbine High School in Littleton, Colorado, ended not with Web page defacements or billions of dollars in damages to online businesses, but with the senseless deaths of 12 other teenagers and a teacher. Nobody saw it coming, least of all the two boys' parents.

The first funeral for the victims at Columbine hadn't even taken place yet when the so-called experts began to blame the Internet for the increase in school violence and the growing disconnection from mainstream society of America's teenagers. Suddenly, any teenager who spent too much time on the Internet and less time taking part in social events or wore black clothes to school immediately became a potential school shooter. America's teenagers were out of control, the experts said. They ran down a laundry list of school shootings to prove their point. All of the kids who had been arrested for school violence during the last few years had similar characteristics. They were loners, geeks who felt persecuted by others. They didn't fit in, at least not with the most popular kids in school or with the jocks.

Of course, there was another group of teens who was believed to share similar characteristics but of which even less was known. This other group even had its own subculture. These kids belonged to a murky underground of criminals and miscreants. This was the group responsible for hacking into corporate computer systems, stealing credit card numbers, and replacing the content of Web pages with vulgar, disturbing, and sometimes hate-filled messages. Suddenly, if you were a teenage hacker, you were a Harris or a Klebold waiting to happen, according to the experts.

But ignorance breeds contempt.

^ () < [] > () *^*

It was around the time of the Columbine massacre, at the same time that word of mass tortures, rapes, and killings began filtering out of Kosovo, that "Genocide" thought seriously about changing his hacker nickname. He thought about it again after the September 11 terrorist attacks that killed more than 2,500 people at the World Trade Center in New York, at the Pentagon outside Washington, D.C., and in Pennsylvania.

But changing your screen name is not that easy. The hacker subculture has a strong undertow. The longer you wait before you try to get out, the deeper you sink into it. Genocide has been hacking since he was 14. That was 12 years ago. Changing his name now would be like committing suicide or faking his own death. A hacker's handle is his identity. It encompasses everything from his reputation in the underground to his capabilities and resume of hacking exploits all in one word or phrase. If Genocide were to change his name now, he would, in effect, become nobody again. He'd be just another wannabe. And Genocide is not a wannabe. He's the real deal.

If you were to ask him why he chose such a horrible word for his hacker handle, a word that produces images in peoples' minds of Nazi death camps and mass grave sites, Genocide would tell you that his decision had nothing to do with expressing support for such acts. Rather, he was making a statement designed to prove to people that they had become immune to being shocked by the horrors of orchestrated murder in the world. People still ask him what the name is all about, but the answer is always the same. A lot of people just don't get it. And if he could change it, he probably would. But he can't. So we should just get over it. Genocide is a fact in the world. Yes, it's an evil fact. It also happens to be the hacker nickname of a kid who grew up in Fairbanks, Alaska. But it doesn't make him Hitler, the butcher of Belgrade, or Osama bin Laden. He's just a hacker.

^ ( ) < [ ] > ( ) *^*

There weren't a lot of computers in Fairbanks when Genocide was a teenager. He certainly didn't have one at home like most kids today do. In fact, he didn't even have electricity, a telephone, or running water in his house. In that respect, Genocide offers a truly unique profile of a teenage hacker.

He grew up in a 20-foot by 20-foot shack in the backwoods of Alaska. It was a single-level house with floors and walls made of plywood, and a plastic vapor barrier on the inside to keep out the moisture. In some places, you could see straight through to the insulation in the walls. Genocide cut wood every day to feed the inefficient wood-burning stove that shielded him, his younger brother, and his mother from the biting cold at night. The family had to take showers, wash their clothes, and make telephone calls in downtown Fairbanks, some 35 or 40 minutes' drive away.

Because of its location in central Alaska, Fairbanks is known as the Golden Heart of Alaska. It sits 358 miles north of Anchorage and only 188 miles south of the Arctic Circle. Fairbanks is a staging point for North Slope villages such as Barrow and the oil fields of Prudhoe Bay. It's a place where summer days last forever and the shortest winter day enjoys less than three hours of sun light. Temperatures in and around the central regions of Alaska range from 65 degrees below zero—a challenge for even the most efficient of wood-burning stoves—to +90 degrees in the summer. It's a pristine setting distinguished by flat, treeless tundra landscapes that are surrounded by the frozen white peaks of kingly mountains that rise up and glimmer beneath the aurora borealis. One look at the surrounding area from the side of the hill where Genocide's house was located and you will see computers and the Internet for what they really are: crude inventions of mortal men.

Genocide's parents had been divorced since he was five years old. His mother raised him and his younger brother alone and relied on food stamps to keep them fed. She had injured her back while working for United Parcel Service (UPS) and was no longer able to earn a paycheck. His mother's strength and will to survive, however, spoke volumes to Genocide and inspired him. She was stern and taught him how a man was supposed to act. A man should be strong, but gentle. A man was supposed to provide for his family. Genocide took these lessons to heart. His father, a pilot for a freight airliner, paid him little attention when he was young and offered an example of what not to do.

As a teenager, Genocide built an extension to the family's house, providing two extra bedrooms for himself and his brother. Nothing came easy. Everything the family had, which wasn't much, they earned or made with their own hands. Teenage life was filled with a series of tough lessons for Genocide.

Wrestling was always a big sport at West Valley High School in Fairbanks. Genocide wrestled in the 160-pound weight class during his freshman year and would move up to the 171-pound class during his junior year. He was 5-feet 11-inches tall, with black hair and dark brown eyes and a quiet demeanor that tempered his muscular, wrestler's physique. When he was 16, he finally shed his braces. After eight years of wearing them, he was relieved to be able to smile without exposing a mouth full of metal. He stayed in shape by competing on the swim team during the off-season. Although wrestling and swimming took up a lot of his time, Genocide also found time for jazz band, choir, and even a little acting in school plays. He particularly remembers playing the role of the phantom in *Phantom of the Opera*.

Schoolwork, on the other hand, was never a big attraction for Genocide. English, math, history—what a waste of time. He had a particular distaste for chemistry. Needless to say, he had nothing to brag about when be brought home his report cards, if he brought them home at all. There was one subject area, however, that provided a rare bright spot on his record: computers. Genocide's discovery of computers marked the first time in his life that he truly wanted to learn.

Nothing sparked his interest the way computers did. There was a mystery to computers that made them irresistible. They were the gateway to the unknown, linked by telephone wires that connected him to the far reaches of the world. Sure, that was part of it. But the other part of the allure came from Genocide's ability to use computers to transcend the rules and his normal group of friends in ways he had never before imagined. He was outside of himself when he was sitting in front of a computer. And there were no real, tangible limits to the digital world like there were in the physical world. There were only challenges to overcome.

Most of Genocide's high school hacking career was spent causing chaos on the school's network. He experimented by playing practical jokes. He put the library computers into endless loops running 30-frame pornographic videos and coded Macintosh screen savers that crashed the machines and couldn't be removed. Genocide had no idea that what he was doing could technically be called hacking. But it was hacking. It was experimentation, exploration. And Genocide was learning.

A year or two earlier, Genocide's cousin Tony—he didn't have a hacker nickname; he was just Tony—taught him valuable lessons in social engineering. This was the art of collecting information from unsuspecting individuals by asking seemingly harmless questions or by pretending to be somebody you're not. Tony was known as sort of a small-time local crime boss—you know: the kid whose name is the first to pop into your mind whenever something is missing or broken. He had taught Genocide at an early age the finer points of phone phreaking—replicating the tone used by the telephone companies to initiate long-distance telephone calls—and how to build the tone-generating blue boxes that had been made famous by the first generation of phone phreakers. Such skills came in handy to a kid who didn't have a telephone at home. Genocide had met his first hacker and gained an intimate understanding of the telephone network before he even owned a computer.

But Genocide quickly abandoned blue boxes when he began noticing that the police were picking off his cousin's friends. One by one, Genocide's phone phreaker associates were disappearing from the scene. The days of the phone phreaks were over, even in the backwoods of Fairbanks, Alaska. The telephone companies were too smart, too technologically advanced for that small-time stuff to go unnoticed.

None of that mattered now. Or did it? He had to figure out a way to pass chemistry in his junior year. He didn't know why the hell he needed chemistry in the first place. You could tell somebody that the initials Li on the periodic table stood for the element lithium, or that little strips of paper turned different colors when you dipped them into a liquid that contained hydrochloric acid. But who really cared? Genocide didn't. Regardless, the class had been dogging him for a while, and if he didn't pass, he wouldn't graduate.

He was in the chemistry lab after school one day making up work, as usual, when his mind started to piece together the answer to his problem. It had been sitting right there in front of him the entire time. It was a Macintosh computer on his teacher's desk. He'd watched her use it before, especially after big tests when she would stay late to enter all of the grades into some sort of database. He realized then that this wasn't a stand-alone system; it was part of the school's network. Now he was looking at something he could understand. A network. Genocide stood at the brink of his first hack, and the pay-off would be huge. He'd get to graduate on time with the rest of his friends.

But he would have to time it just right. When his teacher came in and sat down at the computer, like she always did after the last period of the day, to help the strugglers, like him, he would have to spring into action. He'd have to catch her off-guard, hit her with a flurry of questions to keep her off balance. And that's what he did. She came into the classroom holding a stack of papers from the recent test that the class had taken. None of the other students in the room paid much attention. There was nothing unusual about the teacher's arrival; she arrived at the same time every day. Genocide, sitting at a desk with his head buried in a book, glanced up and watched her closely. If he jumped up too soon, he would blow it. He needed to time his offensive so that he asked his first question as she was entering her password.

If school officials were concerned about the security of the network, they didn't tell Genocide's chemistry teacher. She threw the tests down on the desk and sat down in front of the computer, her back conveniently to the class. Genocide grabbed his notebook and his pencil and walked through the rows of lab tables toward the front of the class. Of course, nobody else in the class suspected anything. Genocide kept information about this operation highly compartmented and protected. He was the only one who knew of his plan, and he wasn't about to start bragging.

The teacher glanced down at the keyboard and prepared to peck away at her password using, as she always did, her two index fingers. This took a lot of concentration, you understand. She was focused. That's when Genocide hit her with a barrage of questions, feigning interest in chemical compounds and equations.

"I can't get these equations to balance," he said. "Is elemental hydrogen diatomic? In single replacement, is one reactant always an element? Does it matter if the element is written first or second on the reactant side of the equation?"

As always, his teacher remained focused. She preferred to do one thing at a time. Genocide stared over the top of his notebook, peering down over her shoulder as she typed. He scribbled in his notebook. But instead of working on the equation, he was actually writing down the teacher's login ID and password to the network. She turned around slowly, with a look on her face that said she hadn't quite heard or understood what he had asked. She was a thinker. Multitasking didn't come naturally to her.

"I'm sorry," she said. "Did you have a question?"

"Yes. I just wanted to know if hydrogen is diatomic," Genocide responded.

"Yes, of course. Remember that," she said.

"I'll try. Thanks."

The next day, Genocide arrived at school early and went into another class-room where he regularly attended speech lessons. There was a Mac in that classroom, too. He jumped on the computer and entered his chemistry teacher's name and password and hit Enter. It was that simple. He found her personal directory on a restricted network drive. Then he scrolled down and found his name and double-clicked it. In less than 30 seconds, he pulled up his entire work history in chemistry. At the top of the list was his grade for the last exam. It was a 63. A click of the mouse, a tap on the Delete key, and bingo: he now had a 73. That was just enough to boost his average for the course into the low D range. Nobody noticed one digit. Graduation was no longer an issue for Genocide. He'd be receiving a diploma next year with everybody else in his class.

^ ( ) < [ ] > ( ) *^*

He wasn't exactly proud of his first semi-hacking exploit, but Genocide never pretended to always play by the rules anyway. That was a good thing, because from that point on he rarely did. His grade-changing caper may not have been what most hackers, including himself, would consider a hack in the true sense of the word, but that was okay. Genocide was just getting started.

At about the same time that Genocide discovered his lust for computers, his mother started taking college courses part time at the University of Alaska. And since they had no running water at their house, Genocide began accompanying her to the college in the evenings to make use of the facilities. The best part about going to the college with his mother, however, was that she had access to the school's network. Of course, his mother had written her login name and password on a piece of paper.

Genocide stole time on the campus network while his mother was in class. The system administrator was a complete poser who didn't know the first thing about securing the system. It was a virgin Unix network run by a virgin administrator whose first rule of thumb was to do no harm. Genocide operated under no such restrictions. He explored the inner workings of the network using his mother's account. He surfed various bulletin boards and spent countless hours reading and learning about various commands, operating systems, and hardware design—you name it, he read it.

Like all true hackers, Genocide quickly became bored with reading about hacking and decided it was time for some hands-on exploration of the university system. He typed:

man [*command*]

A list of commands that could be used on the system filled his screen. He tried commands he was unfamiliar with and explored some more. He tried various logins to see if he could gain system administrator access. He knew that on Unix systems there were a few basic default login-password combinations that lame system administrators, like the one who ran the school lab, rarely ever changed. He tried the ones that he knew:

Root …root

Admin …admin

Sysadmin …sysadmin

Guest …guest

Nothing worked. This wasn't really a big deal. Unix systems didn't record every failed login attempt the way older VAX machines using the virtual memory system, or VMS, operating system did. The school had just upgraded from a VAX system to Unix, so Genocide felt free to explore and learn at will. He was also sort of glad that none of the default logins worked. Breaking in was more fun.

So he told the computer to list the password file for him:

Cat /etc/passwd

Done. To his amazement, the password file wasn't shadowed. That meant it hadn't been replaced by a special token and stored in a separate, unreachable file. To the contrary, it was right there for the taking, although it was encrypted.

That's when he discovered his first version of Crack, a program that decodes user passwords using brute force. Crack was designed to ferret out insecurities in Unix passwords by scanning the contents of a password file and picking out users who had chosen weak passwords, such as common words in the dictionary. These types of users were everywhere on college networks.

Genocide's first successful run of the Crack program was nearly his last. He was still learning and hadn't realized what a system resource hog the Crack program could be. And that could be a problem. Unless the system administrator was completely brain dead, he would be able to tell that something was not right on the network. Genocide allowed the program to continue to run in the background as he opened a new command window and started to spy on the other users in the lab to see if anybody was on to him. The lab was full that night. Nobody seemed to know or care about what he was doing.

At the command prompt, he typed **w**, which told the computer to list all of the users currently online. The list also reported when each user had logged in, what machine the user was sitting at, whether the user had been idle and for how long, and what programs the user was running. That last piece of data was the most important element: with a simple keystroke, Genocide had launched countersurveillance against the system administrator while his password-cracking program ran in the background.

The program needed about 40 more seconds to finish cracking the password file. Genocide took one more look to see what the other network users were up to. That's when he noticed that the system administrator was running a command called a w fstbo. Genocide knew immediately what was happening. The admin had noticed that he was using Crack against the password file. The increase in the server load averages that resulted from Genocide's use of the Crack program must have tipped him off. Maybe this guy wasn't brain dead after all. Genocide also realized that the admin could tell where he was sitting just by looking at the terminal number that was running the program. Genocide freaked out, killed the program and his network session, and ran.

He waited until he was out of the building before he looked back. By that time, another student had already plopped down in front of the computer that Genocide had been using. Computers were hard to come by on campus, and if you got up you lost your terminal in a matter of seconds. This time, Genocide didn't mind. He looked back and saw the admin accusing the student who was now sitting in front of Genocide's system of hacking passwords. Both the admin and the student seemed confused by the fact that the student hadn't even logged on to the network yet.

From that point on, Genocide craved the taste of adrenaline he got from hacking. This was what hacking was all about. The rush and the challenge, the unquenchable thirst for knowledge and the need to push the limits. Hacking was the act of doing something that others said couldn't be done. Hackers solved puzzles that others said couldn't be solved and overcame obstacles that were thought to be insurmountable. Hackers didn't cringe in the face of impossible odds. They became energized when outnumbered. And, most of all, hackers didn't quit. Genocide would eventually get another chance to crack the password file, and he'd be successful. But cracking that password file was just the beginning, and Genocide knew that. For a hacker, learning to crack passwords was like a mechanic learning how to use a wrench.

There would be many other successful hacks and cracks that year. Fairbanks was the perfect proving ground for a young hacker to flex his muscles, spread his wings a little. Genocide was building a portfolio, collecting tools for his hacker toolbox, and gaining a reputation among some of the other regulars in the computer lab. More important, the son of the student had become a student himself. Genocide was a freshman in college studying art and music, and now he had his own authorized access. And he was beginning to make friends, hacker friends.

^ ( ) < [ ] > ( ) *^*

They each knew who the others were, but in the beginning they kept their distance. They were feeling each other out, like a rag-tag army trying to distinguish the officers from the foot soldiers. Genocide had become a regular presence in the computer lab, along with four other hackers. They talked a little, but not much, mostly about security, coding, viruses, and the like. Over time, the hackers became more comfortable with each other. They had struck up a friendship and a feeling of solidarity. Each of the hackers also brought unique skills to the table. Some were better than others at coding or hacking specific operating systems. Gradually, the group realized that they were task organized, with an expert on staff to address any potential challenge.

WiZDom was about five years older than Genocide. He was an ex-Army vet who worked on trucks to make ends meet. At the time, WiZDom was studying for a degree in computer science. He specialized in coding, period. That was what he did best. He wasn't very good with operating system configurations, or anything else for that matter. But a skilled code monkey he was.

Genocide had met Astroboy when he was in high school, but neither knew that the other was into computers. Astroboy was good at working with Macintosh systems, so he instantly became the group's Mac guy and remains so to this day. He, too, was into art, and like most artists had a good imagination.

Alexu was two years older than Genocide and was studying for a degree in music education. Computers were a hobby for Alexu. Years earlier, he had been a regular BBS surfer, and as a result, he had a solid understanding of telephony and old-school Internet hardware design. Alexu also was a gaming nerd in the truest sense of the word. If there was a computer game to be played, this guy had played it. His favorite game was a command-line version of Dungeons and Dragons.

Malcom was the enigma of the group. Nobody was sure exactly how old he was, but from his looks, Malcom couldn't have been more than a year older than Genocide. His background was a well-kept secret. Malcom was by far the best of the group at Linux operating systems. He harbored a weird contempt for graphical user interfaces, though. He insisted on using his laptop running Linux without X Windows (an emulator that offers users the familiar Windows interface). Malcom had also been to the infamous annual hacker conference in Las Vegas known as DefCon before Genocide knew what DefCon was. As a result, Malcom would become Genocide's gateway into that portion of the underground.

A friendly competition emerged within the group, which grew slowly beyond the original 5 members to include about 10 other nameless, ever-changing faces. Somebody would come up with a theory, and it would be up to the rest of the group to either prove or disprove it. They fed off of each other's ideas. Discussions of the theoretical eventually led to outright competitions among the group—virus-writing competitions, to be exact. Genocide, WiZDom, Alexu, Astroboy, and Malcom would each write their own virus in assembly code. Then all of the viruses would be unleashed on the campus network simultaneously. Whoever's virus was left standing at the end of the rampage won.

The members of the group shared similar views about hacking, the inherent freedom of information, and the benefits of knowledge sharing. All were equally pissed off at the ignorance of the media and the general population when it came to understanding what a hacker was. Sure, the five of them had acknowledged during private discussions various "criminal" hacking exploits that they had been involved in, but hackers were not criminals. Those who would censor information and block the pursuit of knowledge were the criminals. Hackers were the defenders of these very basic human rights, according to Genocide and his newfound compatriots. All had agreed on these issues and discovered a sort of intellectual brotherhood within their small gathering. What they had actually done, however, was lay the foundation for what would become the Genocide2600 hacker group.

^ ( ) < [ ] > ( ) *^*

The hacker ethic was all about sharing knowledge. There was no tenet more basic to the hacker community than the freedom of information. All information was created equal and unbound. Knowledge was of no use if it could not be shared. So before they even knew they were a group, Genocide, WiZDom, Alexu, Astroboy, and Malcom formed a local chapter of the 2600 hacker organization and started to share their "skilz"—their skills—with anybody who would listen.

Malcom had come up with the idea for the meetings after he read one of the old issues of 2600 magazine. There were 2600 chapters all over the country, and they all met once a month on Fridays at seven o'clock. Everybody thought it was a great idea and took turns teaching the dozen or so individuals who showed up at the meetings about computer security, telephony, computer media, cryptography, government systems, or whatever their individual specialties had made them uniquely qualified to discuss.

This was a time of great learning for Genocide. He refined his skills in gaining access to password files and cracking them, creating dictionary files for brute-force cracks, exploiting race conditions in server software—that is, taking advantage of another application's use of system resources, such as files, devices, or memory—and injecting instructions into a system by overloading an application with more data than it was designed to handle, to cause what is known as a buffer overflow. He also gained a deeper understanding of human nature, especially as it applied to the way people choose login IDs and passwords. From there, he moved on to some light coding, root kit setup, tactics for erasing log trails, and strategies for becoming an invisible "ghost" on a system.

In addition to being a time of intense learning and technical development for Genocide, this also was a time of rare fortune for him and his family. Genocide and his mother for the first time could afford to wire their home with electricity and a telephone (although to this day running water remains elusive). They paid the telephone company to run the line from a neighbor's house. With the telephone line also came a 75-meghertz Pentium computer that Genocide bought using money from his student loan. Suddenly, the kid who had been living *The Life and Times of Grizzly Adams*—the 1977 television series starring Dan Haggerty as a wilderness hero whose best friend is a grizzly bear named Ben—could boast of having the fastest computer of the group.

The 2600 group continued to hold meetings for more than a year, eventually becoming recognized by the university as an official extracurricular educational organization. Genocide had unofficially taken charge of the group through his own initiative. He was, after all, the tactician who enjoyed reading history and dissecting the battle plans of great military leaders like Napoleon and Frederick the Great. These were timeless lessons from a timeless profession, and Genocide applied them to the hacker's battlefield with great success. This was what made him different from the others in the group. He thought strategically.

After the 2600 meetings, the core group of hackers—Genocide, Alexu, Astroboy, Malcom, and WiZDom-would find a secluded computer room at the university to try out their newfound information or theories on hacking. Their meeting room was located in the same building as the school's auditorium, known as the Great Hall. Outside the room where they held their meetings was a small alcove that housed a soda machine, a candy machine, and a row of pay phones. The candy and soda sustained the hackers and kept the wheels turning through long hours of number crunching and other mind-boggling calculations.

Standard procedure called for the group to dial out from an anonymous number at the university to one of the local ISPs, usually either PolarNet or AlaskaNet, using a hacked Point-to-Point Protocol, or PPP, account. After connecting to the ISP, they proceeded to use telnet—a terminal emulation program that allows a system to connect to a server—to connect to a student account at the local school network, where Genocide had already managed to gain root access. From there, they could do anything they wanted.

Genocide also set up a Web site on the university network and aptly named it Genocide2600—although the directory structure on the university network pointed to an obscure page: http://icecube.acf-lab.alaska.edu/~fstbo. The Web site quickly grew into a major hub of hacking and cracking information. All of the information that was shared by the group at the monthly meetings was at first distributed on floppy disks for the cost of only the disks and then later made available for free on the Genocide2600 Web page.

The original Genocide2600 Web page

In addition to the increasing amount of information on how to take advantage of Unix security flaws, the original Genocide2600 Web site became a repository for dozens of text "philes"—files—about old-school hacking and phreaking. A hacker could visit the Genocide2600 site and learn how to defeat the FBI's lock-in trace capabilities, accept a free long-distance call by lowering the voltage on the phone, make a low-budget two-party phone line, take advantage of a "post pay" public phone—a rural oddity where a caller gets a dial tone first and inserts coins to pay for the call after the other party answers—and make free calls, play tones into a phone and get free credit, tap standard house phone lines, get 12-volt "thief power" from a phone cable, hack and take down a phone company, make a phone busy forever, and construct a chrome box that could switch traffic lights from red to green. These were the building block skills that posers and less experienced hackers chose to ignore. For Genocide, however, studying these old-school tactics was the equivalent of learning the ABCs before trying to read.

Eventually, the university got wind of the type of information that was being stored on the Genocide2600 Web page and asked Genocide to remove it from the university network. So he buried it deeper.

$$*^* * ( ) < [\;] > ( ) *^* *$$

It was around the time when the university started to be a pain in the ass to deal with and the dozen or so pinheads who attended the meetings began to show less and less interest that the real transformation of Genocide and the other hackers took place. This was when they became a real group: Genocide2600.

Of course, in reality the transformation was far less dramatic. As Genocide recalled, one of the group members suggested that the group go over to his house to try some of the tactics and techniques they had talked about at the latest 2600 meeting. The exact words were something like "hey, why don't we run over to my house? I have two machines, a sister, and pizza."

Those simple words changed everything, including the group's social order and the way they interacted with one another. They were no longer faceless peers in a classroom, but people who could measure each other's strengths and weaknesses. As each hacker worked on a particular problem, the others could guess his next move by simply looking over his shoulder. Some of them also began to develop individual techniques but didn't realize it at the time. But a technique can be dangerous. It can act like a fingerprint and lead the police right to your computer. It was because of this that Genocide made a concerted effort to avoid becoming trapped in any one technique.

A few months later, the group held its first off-site meeting at Genocide's house, where they conducted a major break-in. They started scanning the target system, which we'll call "Moon," at 10 P.M. Within an hour, the group had gained root access to the system. At first glance, it looked like a standard corporate server with loads of useless information. It soon became clear to Genocide, however, that the system was actually a major repository of computer security information and hacking tools—not a far stretch of the imagination in an area of Alaska surrounded by two military bases and a university.

Genocide discovered dozens of software tools that were designed to break into computers as a means of testing security. He took them all—about 14 megabytes worth of hacking and cracking tools. Although the "Moon" hack ended the group's use of the PolarNet account, it significantly increased the value and the visibility of the Genocide2600 Web page. The page grew quickly, as did calls by the university to have it removed from the network. So Genocide buried it even deeper.

Genocide2600 was on the map. The tools and scripts the group managed to pilfer were vital to establishing the group as a force and a presence in the underground. "Suddenly we had tools that no one else had, scripts that no one else had seen, and knowledge that no one else possessed on intrusion techniques and new methodologies," Genocide recalled. The heist also was a big score for the hacker movement in general because a lot of the tools that the Genocide2600 crew had managed to get their hands on had been designed to assist the white-hat hacker community: the good guys. These were the software applications that security administrators used to ferret out the bad guys, like Genocide.

About 500 e-mail messages a week poured in from wannabe hackers asking Genocide to help them learn how to hack. Genocide's answer, if he answered at all, was always the same: go learn yourself like the rest of us did.

The increased availability of hacking tools and scripts also helped boost attendance at the group's Friday-night 2600 meetings. But Genocide was careful to warn the other members of the group not to discuss anything illegal. Anybody was free to attend the meetings as long as they took place on school property and were held under the auspices of a university educational program. From time to time, the meetings attracted university officials and professors. And although the stuffed shirts would never publicly condone the methods used to gather the information, they also would never say that they didn't learn something from the meetings.

But it was the presence of one particular stranger at the next meeting that got everybody's attention. This guy wasn't your typical 1960s-holdout university professor. For one thing, he was dressed in a navy blue suit and tie. And he sat in the back of the room, listening to the presentations, emotionless. He seemed to be fixated on Genocide.

As the meeting came to a close, the attendees mingled and began filtering out of the room. This was normal. Presentations often raised more questions than could be answered in the time allotted to the meeting, and the attendees often continued their conversations as they left. The newest visitor to the meetings hadn't left, however. He was still in the back of the room, only now he was standing like everybody else. Genocide looked at him and saw a man with his hands in his trouser pockets, staring at the attendees with a smirk on his face, as if he knew their darkest secrets or had been rifling through the contents of their bedroom dressers only moments earlier.

When the last of the wannabes had left the meeting, the stranger in the back of the room walked up to Genocide and asked Genocide if he would take a walk. Genocide said sure, and the two walked down the hall away from the crowd that still mingled about. That's when the guy in the blue suit showed Genocide his FBI credentials and told him he was leading an investigation that would prove Genocide's involvement in the hack into the "Moon" server. He also ran down some of the evidence he had gathered so far, including the contents of the Genocide2600 Web page. Satisfied that he had put the fear of God into Genocide, the FBI agent simply walked off. The FBI had other tentacles in motion, however, that Genocide would soon learn about.

One of the first things Genocide did was call his mother from the university. He was genuinely scared. But as the thoughts raced through his mind, he realized that maybe this was what the FBI had hoped he would do. As a result, the conversation he was planning to have with his mother took a radical turn. Genocide remembers the conversation as follows:

Mom: Hello?

Genocide: Hi, mom; how's your back?

Mom: Same as usual; do you need a ride home?

Genocide: Naw, I'm fine... Listen, I've got some news...

Mom: Okay—what's wrong?

Genocide's mother always seemed to know from the sound of her sons' voices when they had done something they shouldn't.

Genocide: I think everything is going to be okay, but there are some people here at the school who think I did something, and they are plenty pissed off at me.

Mom: What exactly are we talking about here?

Genocide: It has to do with computers; they think I broke into some high-brow server somewhere and...

Mom: In the school? Did you?

Genocide: No! No!

Mom: Can you prove it wasn't you?

Genocide: It's impossible for me to prove I wasn't there.

Mom: Then it's simple; you just tell them it wasn't you. They will check in the server and find out it wasn't you, and then you'll be on your way.

Genocide: It's not quite that easy mom.

Mom: Sure it is.

Genocide: Mom, it's the FBI.

The mention of those three little letters—FBI—sucked all of the sound out of the telephone. It was as dead and motionless as a black hole in outer space. Genocide could imagine the look on his mother's face at that very moment. She was pissed. Seething. Unable to speak.

Mom: I hope for your sake you didn't fuck yourself. Does your brother know?

Genocide: No, I barely found out. No one knows but you and me right now...well, and the stiff in the suit.

Mom: How do you know it's the FBI and not some trick?

Genocide: Mom, the guy confronted me face to face, showed me his badge, and then told me he was here to prove I broke into that place.

Silence again filled the telephone receiver. Genocide's social engineering skills were never good enough to conceal his own sense of fear from his mother. Not even the best hacker can defeat a mother's ability to detect bullshit.

Genocide: I think I'll be all right.

Mom: I sure as hell hope you'll be. We don't need this right now.

Genocide: I know. I think they may want to talk to you.

Mom: Why me?

Genocide: Because they probably think I'd talk to you about it.

Mom: You mean like now?

Genocide: I guess.

Mom: That's asinine!

Genocide: I thought you might think of it that way.

Mom: Get home.

Genocide: Okay. I'm on my way.

^ ( ) < [ ] > ( ) *^*

Not long after the FBI showed up at the 2600 club meeting, the university froze Genocide's network account. They wanted to find the Genocide2600 Web page and review its contents for links to the hack against the "Moon" server. Although dozens of software programs and text files were found, the authorities could not link Genocide to the hacked PolarNet account. In addition, although the FBI knew that the dial-up connection made during the "Moon" hack was initiated from the university, they couldn't prove who was sitting in front of the computer at the time of the hack. They could prove that Genocide was in possession of classified government data that was not supposed to be in the public domain, but there was no smoking gun that pointed to him as the person responsible for stealing it. Genocide told them that he had downloaded it from the Internet, and that he couldn't remember where he found it. Three weeks and a dozen missed classes later, Genocide's access to the university network was restored—as if the school's system administrator had ever really been capable of keeping him out.

Genocide wasn't home when the FBI showed up at his house and began questioning his mother about his upbringing, personal relationships, and computer use. The questions drilled deep into Genocide's past. The agents were trying hard to build a profile of the young hacker. Fortunately for Genocide, his mother didn't know the first thing about his hacking capabilities. And even if Genocide had told her, she probably wouldn't have known what he was talking about. The FBI also asked about Genocide's relationship with his brother, suspecting that one of them was acting as the mentor and the other as the hacking protégé.

Genocide was indeed the older brother and the mentor. His younger brother looked up to him and was the type of person who could not refuse a challenge. Although Genocide influenced his brother's computer use, he always believed that his younger brother was the more intelligent of the two and a "genius" when it came to math problems. But patience and persistence are important personality traits in a hacker, and Genocide's brother had neither. His brother often quickly grew bored with things, and working on computers went from being an obsession to something he did just to pass the time. Genocide's brother is currently passing the time in a Fairbanks jail for an offense unrelated to hacking.

The 2600 group was never quite the same after the FBI reared its head and crashed the party. The less committed members of the group stopped attending the meetings out of fear of being kicked out of school. Then the university slammed the door on all 2600 group meetings. That was when Genocide and the rest of the core members of the group formed the official Genocide2600 group. They would never again meet in a public place or advertise their meetings, but their dedication to the hacker ethic would remain intact.

^ ( ) < [ ] > ( ) *^*

It was late on a Friday night, and the Genocide2600 group was holding one of its planning sessions. Of course, the discussion repeatedly turned to Genocide's run-in with the pinhead in the blue suit from the FBI. From there, the rant moved on to the university's complete lack of understanding of the hacker lifestyle, the importance of hackers to the Internet security community, and the school newspaper's obsession with the so-called dangerous hacker who'd been hijacking student network accounts. The hacker ethic was under assault. The forces of ignorance were on the offensive.

But that never changed Genocide's way of thinking about hacking. He never said that he abided by the law 100 percent of the time. Sometimes bending the rules was necessary. The hacker community, the government, and the general Internet community were the equivalent of three super-powers engaged in an arms race. If one of those superpowers were allowed to develop offensive hacking tools in secret, like the owners of the "Moon" server, then the balance of power might be tipped in favor of one group. That, in turn, could lead to a dangerous situation, especially if the one group with all of the tools was the government. In addition, banning one group from having access to knowledge about hacking could spark a heated revolt and lead to the creation of more dangerous hacking tools. Hackers, therefore, were a legitimate force and were as important to the balance of power in the Internet security arena as any other group of users, especially the government.

Hacking was all about the pursuit of truth and not allowing one group of people to deny other groups access to the truth. Allowing the truth to be hidden was unacceptable to Genocide. The true crime was not hacking, but the reluctance of others to rip the veil from the sheep's eyes. "We aren't the criminals that need to be put away. We are the ones you should praise," Genocide wrote in his Hacker's Manifesto in 1997. Sure, other groups and individuals had published manifestos, but none were like this one:

The Social Base of the Hacker
The Genocide2600 Manifesto

People generally believe that hackers have malicious intent as a general rule. This, pardon my language, is a crock of shit and obviously the idea/ramblings of the most generally uninformed people on the net. I do admit that "YES" there are those that are out to only destroy, and yes this group does occasionally add to that at a very small percentage (this will be explained later). But for the most part, we are in the pursuit of knowledge. I do not claim to be a 100% law abiding person, nor does the group. Obviously, if you have heard of us, or even after reading this, you will be shaking your head at this point.

People for all time have feared what they do not understand, what they do not know. You don't know us; you don't understand us. Some have labeled us as terrorists, others as criminals. Ok. Sure. Whatever. Go ahead take the criminals and terrorists away that fight for your rights. After you have lost the battle because your soldiers are gone at your own hand, you'll have no one to blame but yourself. We fight with the greatest tools of all, our intellect and courage.

As a whole we believe in a collective good. We believe that people who try to shut out other people, or people who try to censor our actions, language and activities are the people who deserve none of the above. We cling to our most basic civil rights. We also believe in retribution for what is lost. Eye for an eye mentality is spoken here. Take back what is yours. Bottom line is this: Don't [f***] with us. We do [f***] back.

These were dark times for hackers. But Genocide had crafted a rallying call for his group. They had a higher calling, a mission, a cause, and a raison d'être. He recalled the group of young hackers acknowledging privately that what they were doing was wrong, but nobody considered their hacking exploits criminal. Or, well, maybe it was. "Perhaps we knew in the back of our mind but we didn't want to admit it at the time," Genocide recalled many years later. "It didn't seem like the wrong path. It was adventurous. All of us were adventurous. It was like leaving an etched path to find your own way. We were doing things and going places that most people never even dreamed of. It's sort of the same thrill that a trailblazer gets."

It didn't matter that nobody understood them, or that people didn't take the time to even try to understand. The Genocide2600 group decided it would play its small part in the defense of the hacker ethic. And what better place to show people the good side of hacking than in the chat rooms of America Online, where child pornographers still managed to trade their lowly merchandise.

Attacking pedophiles online was not only a good way to spend a few hours during one of the group's Friday-night versions of modern-day LAN parties; it was also a lot of fun. "It's just something I couldn't stomach," Genocide recalled later. "One of those unforgivable things man does to man. I feel the same way about things like women being beaten by men, people being denied education, freedom of speech. This just was a problem where I might be able to make a difference, and honestly, no matter the energy or resources I or the group exhausted, if one child feels the benefit, then it was all worth it."

A program called AOHell, written by another hacker named Da Chronic, was the tool that the group used to hack into private AOL chat rooms in search of child pornographers and zap them with e-mail bombs that would crash their systems right in the middle of their depraved, heinous dealings. Genocide and the rest of the group scanned the chat rooms for pedophile low-lifes, and the AOHell program alerted them to individuals who were discussing child pornography. When they found somebody engaged in this activity, they attacked like a pack of wolves. Within seconds, their prey's Internet connection would be broken. Then the scumbag would reconnect, and wham! The Gencoide2600 hackers hit him again.

Sure, using the AOHell program was technically against the law. But for a hacker not to take action in this situation would be the equivalent of a pedestrian who witnesses an automobile accident and refuses to help the injured for fear of a lawsuit. It would be unconscionable. And real hackers were not unconscionable villains. Genocide was a real hacker.

$$\star \char"5E \star\ ()<[]>()\ \star \char"5E \star$$

By this point, Genocide's hacking had taken on a life of its own, as had the Genocide2600 group. For months, Genocide struggled to figure out how a college degree in art fit into his life. The answer was that it didn't. And so he left Fairbanks without a degree and began peddling his hacker skilz in search of his pot of gold in the computer industry.

The move was a professional success, and a personal disaster. Genocide paid a steep price for the relatively high-paying job he obtained in the computer security industry. Unable to accept the fact that he was leaving Fairbanks, Genocide's mother "disowned him," to use Genocide's words. So he moved to Oregon alone and without the support of the mother who had shown him endless reserves of strength during his years growing up in a small shack on the side of a hill in central Alaska. Genocide looked back on the episode as another of life's tough, unfortunate lessons.

The move to Oregon, however, did not mean the end of the Genocide2600 group. By now, the group was many times larger than its original five members and spanned several states. Genocide hand-picked regional organizers from the pool of new recruits. Most remained out of the public spotlight, anonymous, behind the veil. Genocide took the biggest risk of the group by maintaining the public Web site. But that was okay. After all, Genocide was the organizer, the planner, the tactician, and the leader.

After he settled into his new digs in Oregon, Genocide moved the Genocide2600 Web site to a new Web server in Portland called Aracnet, where the site's traffic really began to skyrocket. Pirated "warez"—wares—and serial numbers were bigger than ever. And so was the group's popularity. Journalists started to call, and the number of e-mails from admiring wannabe hackers looking for training doubled.

Then the flame war started, with a popular antivirus software vendor leading the charge against Genocide's pirated serial number business. Aracnet froze Genocide's Internet account. Genocide faced possibly thousands of counts of software piracy. But when you're as good as Genocide and his hand-picked team of hackers, things have a way of disappearing from servers, of getting up and moving on their own.

Genocide's account at Aracnet was a simple shell account. To keep him out while the software vendor and the FBI amassed evidence of his software piracy, the ISP implanted an asterisk in his encrypted password file. This is what it looked like:

Before:
genocide:2a08ivnO:1001:1001::0:0:Genocide,,,:/home/genocide:/usr/local/bin/bash

After:
genocide:*2a08ivnO:1001:1001::0:0:Genocide,,,:/home/genocide:/usr/local/bin/bash

The addition of the asterisk changed the password to something that could not be guessed because of the unique method Unix uses to perform one-way password encryption. It was simple, but very smart. However, if the administrators at the ISP had been really smart, they would have created a "tar ball"—a Unix command that creates a tape archive, or tar, that combines multiple files into a single, tightly wrapped file, like a ball of tar. Then they could have given it a random name and stored it on a secret server for the feds to access and build their case.

Instead, the system administrators left the account exactly as Genocide had created it. Bad idea. All Genocide and the other members of the Genocide2600 group had to do was find a hackable server, gain root access, unfreeze the account, back out and erase their tracks, and then log back in using the freshly unfrozen account. From there, deleting data was simple.

With the evidence gone, Genocide2600 remained untouchable. Operations continued at their normal pace. An East Coast cell was established under the tutelage of an expert at social engineering. Now the group was national. And not everybody was a hacker in the technical sense of the word. There were people of all ages and professions, some of whom were complete computer novices but who also donated their time and expertise to the group. Specialization took on new meaning.

$$*^**()<[]>()*^*$$

The timing of the voicemail from the FBI agent was uncanny. It had been only three days since Genocide had seen to it that the evidence of the serial number and pirated software operation had been conveniently lost.

It was about 10 o'clock in the morning. Genocide was at work. His pager began to beep, like it always did when he had an urgent message waiting for him on his home answering machine. He called his machine and entered the secret code to retrieve his messages. There was only one. It was from a man who called himself Mr. Jerkins. He said he was from the FBI, and he wanted to meet with Genocide for a talk.

There are only a few telephone calls that a person might receive during a lifetime that will undoubtedly make the hair on the back of the neck stand up. One is a call from a neighbor while you're on vacation informing you that your home caught on fire. Another is a call from your stockbroker apologizing and then saying you should have sold when you had the chance. Another is a call from an FBI agent who wants to get together with you and "talk."

Genocide went into a state of panic and called an emergency meeting of the local Genocide2600 members for later that night. They must have played the tape a dozen times. They nearly wore the tape thin trying to figure out if this Mr. Jerkins was some poser playing a joke. The voice sounded official, but that wasn't proof enough. His choice of words was cool and detached, and he dispensed them with no obvious slip-ups immediately evident to Genocide's social engineering experts. There was only one way to find out for sure. A meeting was set up.

Mr. Jerkins and three other agents showed up at Genocide's apartment the next day. They were five minutes early. They asked if there was someplace they could go to talk. Having lived in Oregon for only three months, the only safe, public place Genocide could think of was the local burger joint down on Glenn Echo Street about a mile away. They all hopped into a late-model Ford Crown Victoria and drove down the road. Nobody talked. Mr. Jerkins drove with all the seriousness of a funeral procession.

Heads turned when the four suits walked into the restaurant flanking a young guy in a black leather jacket and a black shirt that said "Un-natural disaster, can you feel hells laughter?" The agents waited while Genocide ordered—burger, fries, and a coke; nothing fancy. Oh, and a shake. Burgerville made the best shakes in Milwaukie, Oregon.

Jerkins did all the talking. As soon as he opened his mouth, it became clear that the agents were there to pressure Genocide to switch sides. There was no arm twisting involved, but the agents' intent was clear. Genocide had skills they could use—and better to have him on their side than working against them. They threw a brown manila folder on the table, but Mr. Jerkins put his hand on it when Genocide tried to take a look at it. It was, ostensibly, Genocide's FBI file. Jerkins alluded to all of the information they had collected on Genocide while watching him, both online and offline. They told Genocide that they knew where he was heading, even if he wasn't so sure. Federal prison was the way Jerkins put it.

Genocide feigned being the wide-eyed youth that the feds thought he was. Not once, however, did he let on that they were right about him, that they had him figured out. He told them that he would need to think about their "offer" for a day or two.

Almost as soon as the Ford Crown Victoria pulled up at his apartment, Genocide hopped out and slammed the door. He went straight into his apartment and thought about what the agents had just asked of him. That lasted for about a nanosecond. There was no way he was going to do it. Genocide couldn't be rolled that easily. He rifled through his dresser drawers and got out several fake IDs and a phony passport. Then he counted the money he had in his wallet and factored in how much he had in the bank. What he came up with, however, wouldn't get him very far. Running was not an option.

He threw his IDs back in the drawer. A deep sense of relief came over him. Running was for Mitnick, not for Genocide. So he went to his computer and began drafting an e-mail to the other members of Genocide2600. His hands seemed to shake uncontrollably, but he managed it. The e-mail was simple:

> *I've just been offered a job.*
> *In two weeks the Genocide2600 server goes up, the new dawn.*
> *http://www.Genocide2600.com.*
>
> *-Genocide*
> *Head of the Genocide2600 Group*
> **Embrace Freedom**

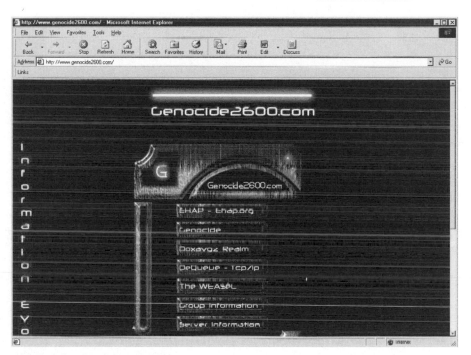

The new Genocide2600 Web site

❄^✷ () < [ ] > () ✷^❄

To this day, Genocide is reluctant to provide more details about the hacking exploits that led the FBI to his doorstep and the many other hacks that the authorities doubtless know nothing about. Like many hackers who have cut their teeth doing not-so-popular and not-so-legal things, Genocide understands the finer points of the statute of limitations.

But there is one current activity that Genocide is happy to talk about: his work with EHAP, Ethical Hackers Against Pedophilia. EHAP is a nonprofit organization composed of hackers and other concerned citizens that use, in the words of the organization's mission statement, "unconventional and legal tactics" to help law enforcement officials track down adults who exploit children online. Genocide has spent the last several years helping EHAP rid the Internet of those who traffic in child pornography. It has been one of the most satisfying aspects of Genocide's hacking career. He has passed information on suspected pedophiles to some of the very same FBI agents who paid him a visit years earlier.

At the age of 26, Genocide bears the scars of an experienced hacker. There are other scars, too. It's been about a year since a short-lived marriage ended in divorce. Likewise, his mother remains a stranger to him. But the memory of that small, wooden shack that sits on the side of a hill in Fairbanks, Alaska, shines bright. It is a symbol of what a person can do, of earning his hacker stripes and rising up from nothing and landing a well-paying job for a major computer hardware and software manufacturer. "I'm just your standard mild-mannered security guy," he says.

Genocide2600 now claims more than 100 members from coast to coast. But the days when hackers had the upper hand because of clueless system administrators are over, acknowledges Genocide. "Today, admins are sharp, weathered hackers themselves, and they're college trained and field tested," he says. "This calls for a whole new approach for hackers. You can't just pick up a script, kick it off, and watch as systems fall to you without getting tracked right back to your front door and busted all in one single movement."

Genocide2600 has certainly evolved since its formation in 1995. Some members of the group are married and have kids. Others are single, mere kids themselves. But the questions about the group's namesake continue. And that brings us to where we started: ignorance breeds contempt.

When it comes to the hacker underground, people choose to see what they want to see and believe what the media feeds them. Genocide would change the name if he could. But there's too much history there. And it was never about hate and racism anyway. The Genocide2600 group is not the band of Internet terrorists that the name might imply to some people.

"The point is, we could be your neighbor or your babysitter for all you know," says Genocide. "We could be the kid filling your gas in your car. It doesn't matter. All you really need to know is that we are spreading as fast as knowledge…at the speed of information."

2

Rebels: Joe Magee and "Noid"

They could have been anybody's kids.

In 1989, two young boys, whose paths would never cross, shared a journey of technological discovery. Both were middle-class white kids who lived in ethnically diverse areas of the country. One lived in the bustling city of Philadelphia, the other in a quiet suburb of Chicago. One was quiet and reserved, the other constantly pushing the limits of authority and testing the patience of the law and losing. For every personality trait they shared, there was another that was all their own. But when the 10- and 12-year-olds discovered their parents' VCRs, they simultaneously embarked upon a journey that would eventually introduce them to computers and take them through the back alleys of the Internet. There, they entered a mysterious, uncharted universe known as the hacker underground.

As teenagers, both boys sank deeper and deeper into the hacker culture, eventually spending countless hours sitting at their computers. They explored the Internet's darkest alleyways and ventured into the unknown, often into places where they shouldn't have been and doing things they shouldn't have been doing. One of the boys confined the bulk of his activities to exploring, like an Internet-age Marco Polo. The other became a cyberspace buccaneer, a pirate interested in cracking and sharing pirated software. But in the end, each boy made a decision that would change his life forever.

The hunger started in 1989, when Joe Magee was 10. It wasn't the typical kind of hunger, though. This was different. It was the kind of hunger that had motivated fifteenth-century explorers like Christopher Columbus to venture

conne
03(*&
1(83)
01*(8
^20=1
01>09
[^(*)
^*201
(*&7]
(83)^
=13+0
099*[
*)flo
()^*2
conne
03(*&
1(83)
0=13+
1*(83
20=13
1>099
^(*)(
*2010
*&7]1
83)^2
13+01
99*[^
)floo
)^*20
onnec
3(*&7
(83)^
=13+0
*(83)
0=13+
>099*
(*)()
20103
&7]1*
3)^20
3+01>
9*[^(
flood
^*201
nnect
(*&71

into the unknown. It was the same hunger that had propelled man to the moon in 1969. This hunger lived in the mind, heart, and soul, and it couldn't be satisfied easily.

The story of Joe Magee's hunger starts in a modest row house in the Irish-Italian section of Southwest Philadelphia. It was a nice neighborhood in those days. All of the homes had colorful, well-manicured gardens in the front and paved driveways in the back. You could walk to nearly anywhere you needed to go or catch a trolley car into the city center. All of Joe's closest friends lived within a mile of his house. It was a good place for a young kid to grow up, even if it didn't stay that way.

It was a warm summer night, about two o'clock in the morning. Young Joe Magee was sitting in his bedroom on the second floor crunching through the possibilities of how his parents' new VCR worked. It was unlike anything he had ever seen before. Somehow the electronic gadgetry inside the box turned two wheels in the cassette, and those wheels turned the tape, which sent a signal through the box and into the television—and bingo: a video played on the screen. Joe wanted desperately to know how this strange device worked. The thought of it had kept him from falling asleep for hours. The engine of his curiosity had been working overtime. It was consuming him, and through his intense concentration he had worked up an appetite.

Most parents expect their 10-year-olds to be in bed after midnight, and the Magees were no exception. Joe's mother, a telephone operator for a local Marriott Hotel, had already gone to bed. His two older sisters had moved out of the house when he was 6 and had started their own families a block away from their parents. But somewhere in a room on the other side of the house sat his father. For the elder Magee, a 28-year veteran firefighter in a city that averaged 100,000 emergency calls and almost 3,000 structure fires every year, the late hours of the evening were a time to sit back and put his feet up. His work schedule was four days on and four days off, and on his days off Magee earned extra money as a carpenter. These were hard-won moments of relaxation and safety for this member of Engine 68, Ladder 13.

Back in his bedroom, Joe Magee's hunger grew by the moment. But he knew he couldn't just stomp into the kitchen and raid the refrigerator. If the old man caught him up so late, it would undoubtedly mean a trip straight back to bed and a long night of wrestling with a growling stomach. But it was worth the risk. So he did his reconnaissance, listening first to the sounds as they bounced off the walls to triangulate the location of his father in relationship to the kitchen. He pictured his own movements in his mind before he made them. Then, slowly, like a cat burglar, he began to make his way out of his room, down the stairs, and toward the kitchen.

When he reached the bottom of the staircase, he turned to his left and looked across the living room. His father was sound asleep, cradled snugly in the plush recliner that faced the television. He tiptoed through the middle of the living room, his father to his right and the television to his left. A late-night edition of Benny Hill threw weird shadows against the walls. He stopped for a brief moment to watch; Benny Hill at two o'clock in the morning was forbidden fruit for a 10-year-old, even if he didn't quite get all of the jokes.

He entered the kitchen undetected and made a hard left turn straight for the refrigerator. He paused to listen. His ears detected no movement yet from his father. He pried the refrigerator door open gently and held down the little white button that controlled the light with his finger. Joe knew where he was going and what he was looking for, so he was in and out quickly: the peanut butter, the jelly, and two slices of white bread. He moved the ingredients over to the kitchen table, out of his father's direct line of vision, and slapped the sandwich together quickly. The deed was done. Now it was just a matter of erasing his tracks and making it back to his room on the second floor unnoticed. He was only a kid, and it was only a peanut butter and jelly sandwich, but the preplanning, the reconnaissance, the preparation to erase the evidence of his movements were skills he could and would use later in life as a computer security and hacking expert. It was good training.

He began creeping toward the stairs, sandwich held tightly in his hand so as not to drop even the smallest bits of crumbs along the way. His eyes were open wide, and his heart rate increased as he moved. Every squeak of the floor threatened to give him away. Joe could see the stairs as he entered the living room. He was almost home free. But something was wrong. Something didn't look right. It was his father. The old man was no longer asleep in the recliner. It was difficult to tell exactly where his father was at that moment or what direction he was moving, but it was clear to Joe that his father was, in fact, on the move and threatening to uncover his clandestine sandwich operation at any moment.

The young boy panicked. Any kid his age would have panicked. He would look back on this episode years later and laugh, but at the moment he was a kid who should have been in bed. And for Joe, this wasn't the first time he had been up when he shouldn't have been. His parents had scolded him several times before for staying up late and making midnight sandwich runs. He was keeping hacker hours even before he knew what a hacker was. The insomnia that would plague him his entire life and become the lifeblood of his hacking activities had already started.

There was no time to make it up the stairs without being noticed, so he ran back into the kitchen. Then it hit him: "I'm probably going to get caught, and it will only be worse if I get caught with this sandwich in my hands." What was he going to do with it now, though? He couldn't put it in his pajama pocket or hide it in his shirt. If his father found him, he would surely find the sandwich, too, and that would spell the end of his plans for a late-night snack. At this point, however, Joe Magee had already lost his appetite and wanted nothing more than simply to hide his sandwich until the morning. There was only one place he could think of where his sandwich would be safe and remain unnoticed overnight: his parent's new VCR.

The Magees had placed the combination television and VCR on a shelf in the kitchen near the back of the house. Joe shoved the sandwich into the slot on the VCR where the videocassettes normally went. The sandwich was a little bigger than the tape opening on the VCR, so he had to give it a few extra pushes with his fingers. The small plastic door that hung in front of the tape opening and protected the internal gears from the damaging effects of fine dust particles closed slowly behind his sandwich. He wiped away some jelly that had rubbed off on the tape drive cover and licked his fingers. Suddenly, a change of heart came over him. This was wrong, he thought to himself. So he pushed two fingers back through the opening and grabbed his sandwich, but the bread pulled apart, leaving the bulk of the sandwich still lodged in the machine. He shoved the one piece he managed to free into his mouth to hide the evidence. A split second later, his father appeared in the doorway of the kitchen.

"What are you doing up so late?" his father asked, a slight hint of aggravation evident in his voice.

"What?" He could feel his father's eyes burning a hole in his chest as the old man stared with that x-ray vision that only parents have as the piece of sandwich made its way down his esophagus and toward his stomach. There was no way he would get out of this one. "I don't know how I got down here, dad. I must have been walking in my sleep again," he said, trying not to lick his lips or stare at the VCR.

His father couldn't argue with him. His son had experienced similar sleeping problems before. "C'mon, let's go back to bed," Joe's father said.

The elder Magee escorted his tired young soldier up the stairs and back to his room and reassured him that he wasn't in trouble. Joe had no choice but to go along with events as they unfolded. He returned to bed and pretended to go to sleep. What he really did, though, was spend the next hour devising a plan to get his sandwich out of the VCR without his parents' finding out. Unknown to him, however, the low-level heat generated by the VCR's electronic circuits had already begun the process of melting the cold jelly in his sandwich. Each passing minute, the sandwich was dispensing a thick, purple, sticky goop throughout the inside of the VCR. There was no denying it now: one way or another, he would be figuring out how that VCR worked.

$$*^* ( ) < [ ] > ( ) *^*$$

Halfway across the country in Skokie, Illinois, a middle-class suburb of Chicago, a father returned home after a long business trip. Like most workaholic salesmen, he spent a lot of time on the road, away from his family, so it was a special occasion when he finally walked through the door to stay awhile.

This time, however, seemed more special than usual. In his hands, he held a large box. His wife, 12-year-old son (we'll call him Aaron), and 9-year-old daughter were intrigued by the thought of what could be inside. Each thought it might be something special for him or her. But then the father carried the box into the family room, placed it on the floor in front of the television, and said, "Guess what I won at work today."

It was a VCR. He had won it as a reward for being selected salesman of the month for the marketing firm where he worked. His wife and daughter were not particularly thrilled, although a new VCR would be nice to have around the house. But Aaron was fascinated immediately. This was cool.

VCRs were not new, but they were new to Aaron. The first crude VHS VCR hit the market in 1977 and had a two-hour recording limit. By the mid 1980s, however, videodiscs had been introduced, along with VCRs with stereo sound, 8mm camcorders, Blockbuster Video stores, and Nintendo Entertainment systems. Aaron also recalled seeing an advertisement earlier in the year for a liquid crystal display, or LCD, video projector. By 1989, though, only about half of all households with televisions also had a VCR.

Aaron had to figure out how the VCR worked. The Apple II computer that his parents had bought for him and his sister a year earlier was cool, too, but it didn't have a modem and was limited in capabilities. As a result, it hadn't really sparked his interest. But this new VCR drove him crazy. And he didn't just want to know how the tape went into the machine. He wanted to know everything, including how the tape moved through the system and how the video signal was picked up by the television.

The VCR proved to be an excellent launching pad for Aaron's hacking and cracking career, which, unknown to him, lurked right around the corner. Although the VCR would not allow him to hack in the true sense of the word, it provided a fertile electronic playground that would help form the basis of his understanding of how all things electronic, especially computers, worked. He started learning by reading the manual. For Aaron, however, reading the technical manual was like reading an Alfred Hitchcock suspense thriller. He couldn't put it down.

What Aaron soon discovered was that a VCR works much the same way as a computer: it converts a signal from the tape into a signal that can be understood by the television and converted into output on a screen. A computer, on the other hand, converts machine language (zeros and ones, bits and bytes) into text and images on the monitor that can be understood by people. Of course, the level of interaction is far greater with a computer. But it is all technology, and through his curiosity Aaron was laying the foundation of his technical prowess. In him, too, the hunger had gotten its first taste.

Even the construction of the videotape gave Aaron invaluable lessons in the logic behind mechanical storage devices. A tape stores its video signals in a series of horizontal line scans, similar to the way a computer hard drive stores data in sectors and blocks in a circular pattern on a hard drive. In addition, a tape has two control tracks that tell the VCR important information about how it needs to play the tape, similar to the way a computer relies on an operating system, software settings, boot files, and other intricate software and hardware interfaces to tell it how to boot and run various applications. Aaron began to see the similarities between electronic systems.

Aaron wanted desperately to get inside the VCR, inside the machine's mind and, in a way, inside the mind of the person who built it. After all, every electronic device inherits a distinct personality from its designer—you can almost trace the electronic DNA of the system to its designer. The eccentric personalities and characteristics tend to stand out more than others. But if you know what you are looking for and look closely enough, you can tell what kind of person the designer was, and whether that person really understood his or her business.

Finding the time and the privacy to take apart the VCR and begin his exploration was never a problem for Aaron: his parents were hardly ever home. His mother, a clothing buyer for a major retail clothing store, shared her husband's drive to climb the corporate ladder and worked late on most days. As a salesman and marketing executive, Aaron's father was rarely home during the week. Instead, he returned on the weekends and then spent most of the day on Sunday watching sports and not wanting to be bothered. A Polish nanny, who lived in a room next to Aaron's in the upstairs addition to the house, took care of Aaron and his sister while their parents were working. The nanny, however, was not his mother. Aaron enjoyed having the run of the place while his parents were gone.

He started early one morning, disconnecting the VCR from the television, unplugging it from the wall outlet, and setting it down on the floor in front of the entertainment center. He prepared his father's tools the way a surgeon or a dentist prepares for surgery. The manual was close by, but he had studied it enough already not to need it. The top came off more easily than he had expected. He was in.

The VCR's innards looked far less complicated than those of his Apple II computer. A few wires, gears, rollers, tracks, and a big cylindrical drum comprised most of the internal component list. He plugged the power cord back into the wall outlet and inserted a tape. Things began to move. Rollers were inserted into slots on the tape, which wrapped tightly around the drum and got pressed against the video and erase heads. This all happened with precise timing and synchronization, like a Bolshoi ballet performance or a dance sequence on Broadway.

Aaron recalled the first time he ever popped the top off of his Apple II computer and looked inside. What he found was nothing less than a work of art, a masterpiece of ingenuity. Steve Wozniak, the father of the Apple system, knew what he was doing when he designed the first Apple system board in the late 1970s. Aaron based all future electronic designs on what he saw that day inside his Apple II. You could see the desire for perfection in the layout of the circuit board. It was simple, but efficient, and designed with expansion in mind. Wozniak's passion lived in every board, like static electricity. The circuit board oozed Wozniak's hacker DNA. It was right there staring you in the face. And it was contagious.

^ ( ) < [ ] > ( ) *^*

Joe Magee's VCR episode was not soon forgotten. He had learned the hard way—with the help of his father—that putting a system together again is much harder than taking it apart. Although the machine never quite worked as well after that, Joe managed to escape with a warning from his parents not to take any other kitchen appliances apart when they weren't home. He was fine with that.

By the time Joe was in the ninth grade at John Bartram High School, Southwest Philadelphia had begun to change. It was what the experts called a neighborhood in transition. Crime gradually increased, and people didn't maintain their homes as nicely any more. Ten percent of the student body at John Bartram dropped out of school that year, according to school statistics. Joe, on the other hand, excelled. But he didn't excel at his schoolwork—he excelled at his hobby and his passion: computers. He became the student that the teachers went to for help when they had problems getting a computer to work the way they wanted it to. He already had a reputation and had earned respect for his unusual ability to navigate his way through software and network devices.

He had graduated from taking apart the family VCR to tinkering with a hand-me-down Apple IIe computer. The shelves in his bedroom that had once stored G.I. Joes were replaced by a long wooden combination workbench and desk that he and his father had built. It made for a perfect computer laboratory. His closet in his bedroom now held more than some of his childhood toys; it was home to a growing pile of model rocket kits and various parts to electronic devices he had collected over the years. The Apple IIe that sat on his disk didn't have a modem, limiting Joe's computer experience to his own desktop, but that didn't last long. Soon, almost half of the money he earned each week delivering newspapers was being spent on computer equipment from the local Radio Shack. That's when he moved up to a TRS-80, which he bought used for about $200.

The TRS-80 (or Tandy Radio Shack 80, the store's home brand) was actually an older computer than the Apple IIe, but peripherals and parts were a hell of a lot cheaper. The TRS-80 also was a capable, solid system and accommodated an internal communications device known as a modem. A modem has the unique capability of turning even the most rudimentary computer system into a powerful global communications device. Joe's modem reached out and grabbed pieces of the digital world, turned them into little electronic 0s and 1s, and then translated them back into their original form and presented the information on his computer screen. It put the world on Joe's desk.

The year was 1992, and Joe was an eager, energetic, and inquisitive 13-going-on-14-year-old hacker in training. He wasn't tall, but at 5 feet 8 inches and 160 pounds, the Irish kid with the brown Marine Corps–style haircut and brown eyes was considered to be a "big dude," and tough. He earned most of that reputation on the football field and wasn't known for bragging about it or throwing his weight around off the field. He was quiet, but sociable and likeable. But when coupled with his size, his quiet demeanor was a source of mild intimidation.

But none of that mattered online. The Internet was the great equalizer. It didn't matter what you looked like or how rich you were. The only things that mattered were your attitude and your desire to learn. Joe had both.

He spent countless hours, mostly late at night, dialing into various bulletin board systems, or BBSs, in New York and California. BBSs act as electronic message centers, where people with similar interests can exchange ideas and information. The Well in the San Francisco Bay area was a great place to find technology discussions. Some of the "leet" (or elite) New York hackers and phone phreakers—individuals who were experts in the art and science of making long-distance telephone calls for free—could be found at The Well. Joe also came across members of the Grateful Dead band at The Well. It was a gigantic online community of people with a wide range of backgrounds and interests.

Everybody in Joe's growing circle of hacker enthusiasts knew about the Masters of Deception (MOD) and the Legion of Doom (LOD). But MOD and LOD were not representative of what was happening throughout the scene in general. These were guys who had built a reputation on crossing the line of acceptable hacker behavior. Sure, a lot of hackers like Joe were beginning to learn about telephone switching systems, but the vast majority drew the line at stealing personal credit histories and sending the telephone infrastructure into a nose dive by changing the routing tables on switches. Joe was there to learn and explore, even if he had to learn from other teenagers who didn't understand that there were limitations to what they should be doing online.

Joe also became a regular on ONIX—the Online Network for Information Exchange—a BBS in King of Prussia, Pennsylvania. But this board was for the general computer user scene. Anybody who was capable of dialing in to a BBS could be found on ONIX. So Joe stuck with the boards in New York and California when he wanted to learn something new about hacking.

Eventually, Joe's BBS surfing became an addiction. His insomnia had grown worse. He lived like a vampire, prowling through the back alleys of the Internet late at night and waking up the next morning to the scalding sunlight of the real world. He told his parents he was doing research for school. A few times he showed his father how the computer worked and how he could chat with people across the country, but the experience had only a minimal impact on the Philadelphia firefighter. "He thought it was great that I was doing it. It was better for me than the television as far as my parents were concerned," recalled Joe. His father was genuinely happy that his son was spending so much time on the computer, Joe recalled. "Then, when the phone bill came, I had to explain to my father that research was very expensive."

Joe's "research" was intense. "I was in a zone," he recalled. "I was taking in so much information. I was totally immersed. If I did go to sleep, I'd just lie in my bed and think about it." Within six months of owning a computer with a modem, Joe went from knowing almost nothing about computers and hacking to knowing everything there was to know.

By the time Joe turned 15, the gang of teenagers known as MOD—Mark Abene (a.k.a. Phiber Optik), Paul Stira (a.k.a. The Scorpion), and Eli Ladopoulos (a.k.a. Acid Phreak)—had already been indicted for their rampage through the New York City telephone network. However, putting an end to the "gang that ruled cyberspace" didn't spell the end of teenage hacking and computer-based exploration of the telephone switching system. In fact, the members of MOD had rarely dealt with the rank and file of the hacker community, unless it was to teach them a lesson about their own destructive capabilities. They thought too highly of themselves to waste their time mentoring lamers and other wannabes. You couldn't even get into the BBSs they used unless you were a bona fide member of the gang and had a password. But there were still plenty of other means of learning. And Joe took advantage of those.

Eventually, Joe began to attend meetings of the local chapter of 2600, a loosely knit nationwide organization of hackers, who actually prefer to describe themselves as private citizens with an uncommon desire to learn about computer security. The group got its name from the 2,600-Hertz frequency required to initiate long-distance telephone calls. The 2,600-Hertz tone held the key to the telephone switching system, and in the early 1970s it was discovered that the tone could be replicated by blowing into a toy whistle that came in a box of Captain Crunch cereal. The 2600 hacker group took its name proudly from these pioneers of the phone phreaker movement.

Joe attended 2600 meetings on the first Friday of every month at the Thirtieth Street Amtrak Station under the Stairwell 7 sign. On some Fridays, only a handful of hackers would be present at the meetings. On other occasions, as many as 25 attended.

Stairwell 7 attracted a wide array of characters. There were a few younger hackers like Joe, but most of the members were 18 or older. You either got along with some people, or you didn't. There were the Goths with the long hair, the tattoos, and the weird clothes, and then there were the handful of system administrators who worked for legitimate businesses. Depending on the phase of the moon and a bunch of other unknown environmental factors, meetings at Stairwell 7 could get a little touchy. This was not like your local support group for people with strange phobias, where each person stands up and introduces himself (or herself) and in return gets a warm, hearty welcome from the rest of the group. To the contrary, people tended to congregate with other members with whom they had something in common.

Information flowed and exchanged hands, however. Learning occurred. It was at Stairwell 7 where Joe had a lot of long, detailed conversations about the Internet in general, which BBSs to check out for certain information, and what new interesting IP—Internet protocol—addresses had been discovered since the last meeting. He also picked up information on where to download hacker files and where to find accounts that could be used to log in to systems. Joe also got his first exposure to the Linux operating system at his monthly meetings at Stairwell 7, though he didn't have a computer at home that was capable of running it.

Although nobody admitted it openly, most of the hackers who attended the Stairwell 7 meetings loved movies like *War Games* (1983) and *Ferris Bueller's Day Off* (1986). Both films starred Matthew Broderick as a teenage technowizard skilled in social engineering (basically talking his way out of trouble and into places he didn't belong) and computer hacking. And in both movies, Broderick's characters were skilled hackers, capable of breaking into the school network and changing their own grades. Could Joe have done that? Of course he could have. But there was a sense of reality that pervaded Joe's world of hacking and kept him from going too far. The movies were set in a world of fiction where things always turned out okay for the hero. As the MOD gang had demonstrated, that wasn't always the case when you hacked into systems for real. In the real world, there were consequences, a reaction for every action.

Other than Joe's insatiable appetite for exploring computers and BBS systems, there was nothing in his background or his daily activities (other than his curiosity and desire to learn) that pointed to a kid who would eventually become involved in hacking the telephone system. In fact, he never really knew that the exploratory trips he was about to take throughout the telephone network would be considered hacking. It was a brave new world that had no boundaries. At the moment, Joe was the only one setting the limits.

^ ( ) < [ ] > ( ) *^*

Aaron didn't believe in setting limits. And throughout his teenage years, that would mean multiple run-ins with the law. There was the Girbaud Jeans shoplifting incident; a "misunderstanding" about stereo equipment at a friend's house; breaking and entering into what he and his friends believed to be an abandoned truck; various late-night break-ins at the local college campus just for the fun of getting chased by the school's rent-a-cop; and the time his next-door neighbor called his mother, accusing him of throwing prophylactics at his house. He had actually thrown a condom filled with water out of his bedroom window, but he didn't mean for it to land on his neighbor's patio.

Fourteen was a pivotal, transforming age for the teenager of Turkish and Irish descent. After his parents divorced, his father moved out of the house, and his mother started to work even longer hours than before. But in their final act as a couple, Aaron's parents had done something that would change his life: they bought his sister a new computer.

Helping his sister with her new computer sounded like a good idea to Aaron. If nothing else, he had a new electronic device to play with and explore. He was always game for toying with a new gadget. This time, however, it was more than just a new contraption that the family hardly ever used. This wasn't a new VCR. It was an Apple Macintosh Performa series computer, fully equipped with software and a modem. He taught his sister what he knew from his Apple II experience, but there was so much more to this system. This was different. There was more to learn here than he could imagine, especially when it came to the modem.

"The next thing you know, I'm on the thing until four o'clock in the morning every night just trying to figure out how it worked," he recalled years later.

Soon, Aaron was using the Mac more than his sister was. Gradually, he realized that an entire digital universe, far removed from the physical world, existed on the phone lines. There weren't only other computers out there; there were other people sitting in front of those computers. And they were interesting people from all walks of life. Some of them were mysterious and lived in the shadows, but you could interact with them without feeling threatened.

This was the biggest VCR Aaron had ever imagined. And best of all, it talked back.

But nobody used real names on the BBS. Instead, people used screen names, electronic aliases, or digital nicknames. Aaron needed a nickname.

"Noid" seemed like a natural screen name for Aaron. It was a nickname given to him by his friends and stemmed from his part-time gigs as a DJ for local parties. He'd get so pissed off at competing DJs who talked shit about him that his friends would have to stop him from throwing punches. That's when somebody joked, "You better avoid the Noid." His friends laughed, picturing Noid, the small, red, big-eared, buck-toothed claymation figure employed by the Domino's Pizza company for the lead role in its television commercials. The Noid would jump through a person's window on a pogo stick and try to destroy the person's (non-Domino's) pizza, similar to the way Aaron's friends pictured him jumping all over a rival DJ. Although Aaron had thick, black hair, dark brown eyes, and dark skin and looked nothing like the Domino's Pizza Noid, he was now known exclusively as DJ Noid. Online, it was simply Noid.

By 16, Aaron had developed an alternate existence online. The friends Aaron socialized with at school and around town were not the friends he was making online. And that was cool. Having two sets of friends was manageable and, as he would soon realize, almost a requirement of the digital underground. The cyber and the physical worlds didn't mix. If you weren't a member of the underground, you just didn't get it. And there was no way to explain the addiction to the above-ground dwellers that could make them understand. Regardless of what you did, your nonhacker friends would get on your case constantly about not being able to get through to you on the telephone. This was reason enough for Aaron to keep the two worlds separated.

In the online world, Noid was a Mac guy. That meant he frequented the Macintosh BBS scene. Through his work as a DJ and a dollar-for-dollar matching gift from his parents, he saved enough money to buy his own Mac. That meant relinquishing control of his sister's, but his own computer was installed in the privacy of his bedroom upstairs. Although there was no door to his room, he could hear his mother approaching in plenty of time to hide what was on the screen. Often, he would keep his video games running on his television so he could simply hit the power button on his monitor and pretend to be playing games when his mother came walking up the stairs. She had no idea what it was that he did on the computer. All she knew was that the kid who couldn't leave the house without getting into trouble or getting arrested was now spending more and more time in his room at his computer. This was a good thing.

In reality, however, Aaron was busy learning the finer points of software cracking. He had downloaded a few automated tools from a BBS and started experimenting with his own applications. He ran scripts that could crack authorization codes on copyrighted software. Some of the better tools recorded every instance where an application interacted with the system registry and deduced where the anti-piracy protections were located. Aaron could take that information and, like a surgeon who peels away layers of skin to get to the important organs, pick through the code and remove the protections. Gradually he built a collection of cracked software applications that carried his brand name.

Eventually he took his interest in cracking and hacking, as well as his newfound "skilz"—hacker lingo for skills—to school. At first, whenever he touched a computer at school, it was to play a practical joke. For example, he routinely rewrote the executable boot files on the computers in the school library so funny pictures (funny being a very subjective concept) would be displayed on the screen whenever a teacher turned on the power. On another occasion, he hacked his way into a system and inserted sound files that would launch without warning into an orchestral hymn. The library was never the same once it started to offer students computer access. It certainly was less quiet.

It was all pretty innocent, teenage fun. But Aaron started to earn a reputation for being the kid who could do just about anything with a computer. If the day came when a company invented a computer with legs, Aaron would be the one to teach it to dance. He made new friends in school as a result of this. Kids who hadn't paid much attention to him in the past were paying attention to him now. They egged him on, challenging him to hack into the school computer lab. He did it with ease. Anything that wasn't locked down, Aaron compromised. Anything that school officials said students couldn't do on the computers, Aaron did. Anything that was put off limits, Aaron brought back within the limits.

It wasn't long before the school bestowed a sense of legitimacy on Aaron's computer prowess. The head of the school's computer lab called him in one day and asked him for help in analyzing a series of mysterious system glitches—glitches that Aaron had undoubtedly had a hand in creating. The fact that his recommendations often led directly to the source of the problem only increased his stature in the eyes of his teachers. There it was: they had been staring at the problem the entire time, and in walks this kid and spots it in five minutes. This is real talent, they thought. None of the teachers ever questioned his uncanny ability to pinpoint the problems. Of course, he had the skilz to do so, but none of the adults at the school ever realized that he had the run of the network.

Along with Aaron's reputation as a computer genius came respect. After all, in many ways hacking is nothing but a big ego trip. As with Joe Magee in Philadelphia, Aaron's hunger for discovery and accomplishment fed off of the respect and admiration he was receiving from his peers. He was out to prove that he was smart and that he could defeat any challenge put to him. At no point, however, did his alter ego Noid take over his true identity. Aaron and Noid did not compete with each other for attention. Noid didn't stick out like a sore thumb in school. Aaron saw to that. He went to parties, played sports, hung out with friends. He was sociable and well liked. But when you got right down to it, Aaron spent more time at home, in the privacy of his bedroom, on his computer, being Noid.

He was hooked. He lived and breathed his online experience. Sure, it was only the BBS scene at first, but he was starting to get good at cracking software. Aaron's latest exploit was a crack of Macintosh remote-access server software. This was his best work to date. Until then, he had spent his time sharing pirated copies of games and breaking the authorization codes for relatively harmless software. He had been monitoring and reading the hacking and cracking BBSs for a while, but Aaron was growing impatient and was eager to share his skilz with the online crowd he was running with. He wanted in. And he was about to find the perfect venue, and a mentor to boot.

^ ( ) < [ ] > ( ) *^*

Although Joe Magee was "definitely one of the nerds" in school, he was not your typical pocket-protector type of nerd. To the contrary, Joe was an athlete. He attended football practice after school and played street hockey. He was a big, physical player and a team leader. In street hockey, he was a high scorer and a team captain. Like most kids in the Philadelphia area, he played street hockey most of the year. It, too, became a love affair for Joe, who went on to play ice hockey in college and continues to don his skates to this day.

But from about eight o'clock in the evening until he could defeat his insomnia and fall asleep, Joe rarely left his computer. It became what he later described as an addiction. His girlfriend started to feel ignored and demanded he spend more time with her. So he adjusted his schedule and started his late-night hacking routine a little later. He still managed to stay ruthlessly focused on what he was doing. The addiction fed his insomnia because there always was something new to learn. He never reached an end-state, a final goal. There was no finality to it. The end came only when you ran out of ideas and your imagination dried up.

However, the Internet went through an immense explosion during the years 1991 through 1997, Joe's prime teenage years. There was more to learn during this period than at any other time in the Net's development. By the time Joe was 16, there were more than 3 million hosts and 3,000 Web sties on the Internet. By the time he was 17, the number of hosts had doubled, and the total number of Web sites had surpassed 25,000. Things were changing rapidly. The desire for discovery among the Internet's young pioneers spread like a wildfire on a dry, windy California afternoon. Computers were in homes in record numbers, and at Joe's school, "computers were everywhere."

At school, Joe stood out above his peers in terms of his computer knowledge and skills. He gained a lot of that knowledge, he says, by watching and learning from other kids who were very good at experimenting. "You knew who they were," he recalled. But the teachers knew who Joe was, too, and they quickly recognized the student with the C+ grade point average as the "computer expert" in the school of 2,000 students.

Among his friends at John Bartram High, Joe quickly became known as the "computer dude." Friends and teachers bugged him constantly for help, oftentimes asking him to do things like crack passwords for them or find information stored in obscure areas of the school network. What made requests like these so annoying, however, was that most of the time the kids who asked him for help didn't care how he did it, just that he could. Joe, on the other hand, viewed hacking as an art form, something to be respected and a talent that took as much hard work to perfect as the most challenging school science project. On most occasions, Joe would simply say no to requests for such help. At other times, he would accept the challenge to break into a system for somebody only to pretend later that it was impossible. "Sorry dude. I can't crack this one," he'd say, laughing quietly to himself.

There was one case, however, when a direct challenge got the best of Joe. When a few of his fellow computer enthusiasts directly challenged his ability, Joe gave in and allowed his pride to take over. There were chicks in the room. Need more be said? He wrote a short script on the fly and then logged in to the school network. He launched the script, which disabled every account in the school, including his own so as not to leave behind a smoking gun. He had proved his power to the students who had challenged him, but in doing so he had sent a few teachers scrambling to figure out what was wrong with the network. When the school day ended, Joe stayed behind and went to see his computer teacher. He was one of the few students who had been given administrator access to the network to help the teacher—a fact lost on those who had challenged him: Joe was an insider. He offered to help his computer teacher, Mr. Phelps, figure out what was wrong. To the teacher's amazement, Joe coaxed the network back to life and had it running in a few minutes.

Joe's computer prowess even landed him a part-time gig creating electronic forms for the school principal's personal business. This gave him dedicated access to one of the school's many computers, most of which were TRS-80s. It also gave him a chance to monitor his favorite BBSs from school. And that's exactly what Joe did. One day, he noticed a modem jack in the back of the computer he had been using to help the principal create business forms, so he ran a telephone line from a nearby fax machine to the modem. Now he was sitting at a live BBS terminal, and it was free. The forms that the principal had asked him to create were simple, and he was able to finish them quickly. But he pretended otherwise and used his spare time to surf the bulletin boards and improve his hacking knowledge from school. What else was school for, if not learning? And if he didn't get a chance to check the BBS then, he would go home during his lunch hour and jump on his home system.

Although they were growing in number, BBSs for most of their early history were good for little more than playing games, chatting with other users, and running up high long-distance telephone bills. Computer enthusiasts ran most of the BBSs as a hobby, similar to the Ham radio operator community (only without the license requirement). However, throughout the 1980s and into Joe's era in the early 1990s, BBSs evolved to include rich download areas that offered users a variety of tips and software. One of the most popular topics of discussion on the BBSs in those early days was how to dial into the telephone switching system. Joe monitored those discussions, studied what other hackers were doing, and downloaded text files that contained detailed information and commands for gaining access to the telephone switching system.

The early 1990s were the equivalent of the Wild West for the Internet security scene. Few people, both inside and outside corporate America, gave Internet or computer security a second thought. Although Internet outlaws like Kevin Mitnick were known to be on the loose and news of the crackdown on the MOD gang had filtered out, security controls and the notion of hacking as a crime were concepts that were still alien to most companies. Although the telephone companies were taking steps to increase security in the aftermath of the MOD crackdown, they could not move fast enough. In fact, during his adventures through the back streets of the Net, Joe discovered that very few systems and sensitive data files were protected. And, as Joe would soon learn, this was especially true of the telephone network and the switching systems that controlled it.

﹡^﹡ ( ) ﹤ [ ] ﹥ ( ) ﹡^﹡

Aaron became hooked on a BBS known as Revenge. The board's operator, who went by the cyber handle Apocalypse, was heavily into shareware and freeware. Despite the BBS operator's nickname, users were welcomed to the BBS by a legitimate, professional storefront interface, with dozens of categories from which to choose. Aaron downloaded scores of games and other applications for his Mac. Revenge became his regular hangout online.

Eventually, he and Apocalypse started to chat in real time, exchanging information on software and new cracking tools. Over time, they became comfortable talking to each other—but not too comfortable. Apocalypse was cautious. He had a few secrets to hide, and he knew the feds were beginning to take an interest in electronic crimes. But Aaron seemed to have mastered the language of the underground. He was fast with his responses and hardly ever missed a keystroke or said anything that raised a red flag. He wasn't a goof, and nothing he ever did labeled him as "a fed." Even if he had missed a few keystrokes or hadn't known something that a hacker should know, a federal agent could never have made up stories of run-ins with the law like Aaron's.

Late one night, while he was surfing around the Revenge download area, an idea flashed in Aaron's mind like a thousand-watt bulb. This is a gold mine. Does this guy realize what he has here? Could he be that much of a lamer not to know the size of the market for some of the software he's got here? Actually, it wasn't really the software that Revenge already offered for downloading that Aaron was thinking about, but the software he knew Revenge had the potential to offer, and at a premium price. This was the venue he had been looking for to share his "warez"—his wares, his stuff. So he broached the subject to Apocalypse. He had to be careful, though, because trust was everything in this new world. If he blew it with Apocalypse, he might become the target of Apocalypse and his gang of hackers.

NOID: YOU INTERESTED IN SOME PHILES [files]?

APOCALYPSE: WHAT TYPE OF PHILES?

NOID: THE FREE KIND.

APOCALYPSE: THAT'S ILLEGAL. WHAT, ARE YOU A FED?

NOID: I'M NO FED. BUT HAVE YOU CONSIDERED IT?

APOCALYPSE: WHAT MAKES YOU THINK I HAVEN'T CONSIDERED IT?

NOID: MAYBE YOU HAVE. I DON'T KNOW. HAVE YOU?

APOCALYPSE: HOW DO I KNOW YOU'RE NOT A FED?

NOID: BECAUSE A FED WOULD NEVER DO THE CRIME.

APOCALYPSE: PROVE IT.

At this point, Aaron was faced with a decision. He could upload what he considered to be good crack—the remote-access server software he had recently managed to crack—and face the possibility that Revenge was not who he said he was, or he could back off and let it go. Anybody who knew Aaron would not be surprised by what he did next.

NOID: YOU GOT IT. HOW'S THAT LOOK?

He uploaded the cracked software. No response. A few more minutes, and still no response from Apocalypse. Aaron looked at the power button on his computer and considered shutting down, killing the connection abruptly. A few more minutes passed with no response. This was definitely not "k001," or "kewl," or however you wanted to spell it. For Aaron, the kid who turned to the computer to stay out of trouble, the same Aaron who couldn't afford another run-in with the law, the silence on the other end of the telephone line was deafening. It would really suck if he got busted for this, too. Then, finally, an answer came back.

APOCALYPSE: NOT BAD. NOW CHECK THIS OUT!

In seconds, Aaron's chat window filled with dozens, maybe a hundred, different file directories pointing to cracked software of every conceivable kind. You name it, and this guy had it. There were games, business applications that sold for hundreds of dollars on the commercial market, scripting languages, phreaking and cracking utilities, BBS software that would eventually allow Aaron to run his own board, and even operating system code (especially from Microsoft) that hadn't yet been released to the public. This was serious business, and it blew Aaron's mind. All at once it hit him: Apocalypse had taken him in, had accepted him into the fold. This was his entry into the big time. He was no longer one of the small-time game traders. Now he had access to serious crackers.

In the mid 1990s, software piracy was a plague that crippled the commercial software industry. From 1995 to 1996, for example, the number of pirated business applications increased by 20 percent, to more than 225 million. The cost to private companies around the world exceeded $13 billion, according to the Software Publisher's Association (SPA). In the U.S. alone, the damages exceeded $2 billion. Aaron was keenly aware of the SPA and the effort that was being made to crack down on pirates and crackers. But he felt safe with Apocalypse. The Revenge board was thought to be the third-largest BBS for Mac users in the U.S. Although it masqueraded as a legitimate shareware and freeware storefront, Revenge also offered hundreds of megabytes' worth, maybe even a terabyte's worth, of pirated software in hidden directories that only trusted crackers were given access to.

Aaron began to upload volumes of crack, as pirated software became known. If he had something that he didn't see on the board, he cracked it and uploaded it for others to share freely and, of course, illegally. But Aaron had a few things to learn, particularly about how the board was managed, before he would be granted unfettered access to the inner sanctum of the Revenge board. First of all, it was going to cost a little money to get good ratios. That was the tight-ass, business suit way of saying that the more money you coughed up, the more access you would get. At first, Apocalypse gave Aaron only limited access to the board. But for a "donation" of about $100 a year, he received the keys to the kingdom—everything. Of course, the money was not for the software; Apocalypse was not in the illegal business of selling copyrighted software. Rather, the donation was to cover Apocalypse's expenses in running the bulletin board. Whatever people like Aaron uploaded and downloaded was not his business.

But, of course, it was his business. Apocalypse sat atop the shadowy world of the software cracker. Revenge was more than a BBS; it was like a Mafia organization. It had tentacles that ran far and wide throughout both the hacker and cracker communities, as well as the legitimate software industry. There were those whom Apocalypse trusted more than others. He had moles in software development companies—insiders—who would get beta (pre-release) copies of major software applications for uploading to the board. He employed, so to speak, a cadre of software testers who had access to different applications. The better cracking tools came out of the computer companies that developed the applications—again, insiders. They were, as far as Aaron could tell, the most important part of the scene.

Apocalypse mentored Aaron. "The pioneer who cracks something for the first time, he's brilliant," Apocalypse would tell him. "But after that, everybody is just downloading and using his work." Copycats. Anybody can download a script and crack a basic program. He coaxed and encouraged Aaron to crack new ground, to do the difficult, if not the impossible. Aaron's remote-access server software crack was an example, he said.

Eventually, Apocalypse introduced the young start-up to his crew. But these weren't kids; these guys were real pros, and most of them weren't teenagers. The level of technical competence of some of the more active board members was astounding. A few of the board members had managed to crack a type of hardware encryption system known as dongle encryption. A dongle is a device designed to prevent the unauthorized use of hardware or software. It usually consists of a small cord attached to a device or key that secures the hardware. These were not easy things to crack. If you cracked this type of encryption protection, you knew what you were doing, and you were not to be taken lightly.

Aaron's adrenaline had spiked, and he was staying up late almost every night, uploading his own software cracks when he could and downloading others when he wanted or needed something. He snagged his sister a copy of Adobe Photoshop, a software application used to create graphics and manipulate digital photos. "It's a good program," he told her, "but it costs about $600. Who can afford that?"

Then, to Aaron's surprise, one night Apocalypse confided his location. After all this time, the SYSOP—the system operator—of the biggest piracy BBS Aaron had ever seen lived only a few miles away. That was a little crazy, a little hard to imagine, but true.

^ ( ) < [ ] > ( ) *^*

Within two years of getting his modem-equipped TRS-80, Joe began writing scripts that instructed his computer to keep making telephone calls until it found a number that allowed it to connect, or "talk," to another computer's modem. The process was called war dialing, and it got its name from the movie *War Games*. Joe eventually got a modem for his Apple IIe and tried war dialing with that system, but Apples were not yet widely accepted clients on the Internet, and the systems he connected to spit back crazy characters instead of logical prompts. He connected to a few systems with the Apple, but because of the funky responses, he never knew whether he had reached somebody's personal fax machine or a major node worth exploring on the telephone network. Using the Apple was a pain in the ass, he decided.

The TRS-80, however, solved that problem for Joe. Part of the beauty of the TRS-80 was its simplicity. You could write simple BASIC code that addressed the modem directly. War dialing with the TRS-80, therefore, was a matter of writing a few simple lines of code in BASIC, testing it on a few numbers, and then copying that script to address all of the numbers you wanted to call. Sooner or later, you were bound to come across a telephone switch.

Telephone switching systems provide routing, call supervision, customer IDs, phone numbers, and other services to local telephone users. They are normally located in centralized offices and are a key element in the process that allows consumers to make telephone calls that can be routed throughout the world. All calls pass through a central-office switching system. The copper wires that connect to each individual phone are linked to a system that routes each call. The call gets routed, either within the same switch or to another switch in the same central office, to a nearby switch in the caller's local area or to the toll network. At the destination, the call goes through another central office and then gets routed to the destination telephone that matches the number being called.

Switches are not like computers, however. They are touchy. There is no help or graphical interface to walk a person through the process of logging in and navigating the system. Switches just answer the phone and sort of wait there for the user to do all the work and tell them what to do. A person has to have some idea of what he or she is doing to get beyond the login.

The first time Joe dialed in to a telephone switch, he didn't know it was a switch until 10 days later. There were no banners, no warning messages, no passwords, nothing that told him he wasn't supposed to be doing what he was doing. Once he realized what sort of system he had stumbled upon, however, he "quickly wised up," he recalled. At first, he was afraid to go further. He remembered what had happened to the MOD gang. But then he did it anyway. He wrote down everything he did on a sheet of paper, recording every command and the exact sequence of his actions. Joe not only wanted to be able to do this again, but he also wanted to have an idea of the steps he took in case things went terribly wrong and he needed to tell somebody how it happened.

With commands he had learned from his BBS discussions, he was able to reset his home telephone line. He didn't know he was able to do this, of course, until after the third time he accidentally disconnected the same line he was using to dial in to the switch.

The telephone systems in the United States and Canada handle more than 7 million toll calls a day. The calls are routed across a network of more than 100,000 long-haul trunks that connect approximately 2,600 toll switching offices and pretty much all of the telephones in both countries. A hacker who gains access to the switches enjoys untold powers to wreak havoc. Switching systems mattered, and still do, as telephone users found out in 1991 when two software crashes at separate switching stations that year knocked out telephone service to thousands of homes in Washington, D.C., Los Angeles, New York City, Pittsburgh, San Francisco, and three metropolitan airports. The seriousness of the situation was not lost on young Joe Magee, either. He knew what he had access to and what he could do. With "the commands that I had, I knew I could take out a trunk of networks," he recalled later. "You're talking about the entire Philadelphia area. You easily could shut down the network."

Like the true hacker that he was, Joe called his local telephone company and explained how lax the company's security was. He told the people he talked to how easily anyone with a computer could knock out part of Philadelphia's telecommunications network. But nobody seemed to care. Finally, as his call was routed around to different people at the company, one woman laughed at Joe's story and told him "not to be silly."

The sense of accomplishment Joe got from his telephone switching exploits was unmatched by anything else he had ever done, including taking apart his parents' VCR. "I think I set high expectations in my head of what I could do and then wouldn't stop spending time on the challenge until I beat it," recalled Joe years later. "I always realized that a human being had built what I was attempting to defeat, whether it be a network, a cell phone, or a computer." However, once he realized the power he had at his fingertips, he stopped. It didn't take a degree in rocket science for him to figure out that what he was doing was wrong and dangerous.

Joe could remember where he was on Martin Luther King, Jr., Day in 1990, the way his parents could remember where they were when they heard that President John F. Kennedy had been shot. After all, January 15, 1990, was the day the MOD gang had used a series of bad commands to poison the routing tables that controlled the AT&T long-distance switching system. That was the day when a few kids sitting at their computers had blocked about 75 million telephone calls from taking place. That was the day when a so-called practical joke went horribly wrong, especially if you were a heart-attack victim trying to dial 911. This was neither a game nor a victimless prank.

In the end, Joe's sense of right and wrong prevented him from repeating the mistakes of the MOD gang. But that didn't erase the sense of gratification and accomplishment he got from his exploratory incursions into the telephone system. "I was really proud of myself because it was okay if you were dialed in to the switching system because it wasn't protected," he recalled. "If you had a modem, you had root access to the phone system. That's where I drew the line though."

^ ( ) < [ ] > ( ) *^*

Aaron got a lot of satisfaction from his software cracking experience. He began to earn a reputation. It was intoxicating for a teenager to have his name—in this case, Noid—recognized and respected by others on the board. There was also a high level of entertainment value to what he was doing. Sure, the time he spent at his computer was keeping him from having to call his father to come pick him up at the police station. But figuring out how to crack an application was fun, too. It offered more intellectual stimulation than anything he learned in school. And the challenge was more enticing than anything he had ever faced—it was certainly more challenging than breaking into the local university.

So you can imagine his frustration level when he could no longer connect to Revenge. It was late, and he had just been on the board the night before. Everything had seemed normal. He checked his own system just to be sure; it was definitely the board. He surfed around a bit—nothing. Revenge wasn't even on the Internet Relay Chat, or IRC, channels, the channels that allow users to converse in real time. Aaron stayed up a little longer, trying to connect to the board, hoping it was a temporary glitch. He never made the connection that night.

The problem was of such import that the next day Aaron decided it was time to pay Apocalypse a visit in person. He waited until later in the day—hackers don't operate well in the morning—and walked over to the address Apocalypse had given him. He knocked on the door; it was an apartment. Nobody answered, but Aaron thought he heard somebody inside. He knocked again. Nothing. Then, finally, he could hear the locks on the door turning. The door opened slowly.

"What's up dude? I'm Noid. I've been trying to get to the board all night. What's wrong?"

Apocalypse didn't want to talk at the door, so he invited Aaron inside.

Aaron walked in slowly, and Apocalypse closed the door behind him. Apocalypse was about 27 years old, and he lived with his girlfriend in a ratty little apartment. There were clothes thrown all over the place, ashtrays with dozens of half-smoked cigarettes in them, and tables covered with empty soda cans and beer cans. In the middle of all of this chaos, there must have been five computers—Aaron counted at least that many. Computers were like end tables in this place, they were part of the furniture, the natural decor of the apartment.

"Nice place dude. So what's going on?" Aaron asked.

"There's a mole on the board; the feds are onto me," Apocalypse said. "I'm shutting it down." Apocalypse was either really spooked or completely strung out on caffeine. Regardless, he was full of nervous energy.

"So what are you going to do, really?" Aaron asked in disbelief.

"I told you, I'm done. You can run your own board if you want, but I wouldn't recommend it. Either way, I'm out of here."

This wasn't the first time, and it wouldn't be the last time, that the FBI had infiltrated the underground and spooked a bunch of would-be criminals (and in many cases, experienced criminals) out of the hacking life for good. Technically, what the hackers were doing was criminal. But nobody was into hurting other people. This was a victimless crime. Big, powerful software companies that tried to horde knowledge and, even worse, make money from what should be freely available to everybody couldn't possibly be considered victims. Victimization was for the weak and the innocent. Software companies like Microsoft and Apple were neither weak nor innocent, in Aaron's mind.

The fact that Apocalypse was completely paranoid about the situation made Aaron paranoid, too. Now he was spooked. And Aaron knew he had good reason to be. He knew about the SPA campaign to get the FBI to ferret out crackers. Once in a while, everybody got paranoid about that.

But there was no hard evidence pointing to an FBI mole on the BBS. Of course, there didn't have to be. Just thinking that there might be was enough to scare most reasonable hackers and crackers, who liked to think that they had a professional future working with computers. In reality, word of the board was probably spread by a bunch of "lusers"—losers—who gave out too much information about it on open IRC channels. It had certainly happened before to other piracy boards, most recently to a board called CYNOSURE. Some kid ran that board out of the Massachusetts Institute of Technology. Once it grew to about 200 megabytes of warez, things got out of control, and its members couldn't keep their mouths shut. Eventually, the FBI caught on, and the SYSOP ended up in court. Despite the fact that the charges eventually were dropped against the operator of CYNOSURE, Revenge and Aaron were convinced they were now part of a similar bust.

Aaron went home that evening with a lot on his mind and a major decision to make. He was older now and had a lot less tolerance for this sort of crap. He knew in his heart that he couldn't keep doing this forever. Pictures of FBI agents busting down the front door to his house, pushing his mother out of the way, and running up the stairs to his room to seize his computer for evidence in his upcoming trial haunted him. He would either quit on his own, or he would be forced to quit through a long stay in a jail cell.

When it came time to make a decision about his future, Aaron chose college. He knew he needed to expand his horizons. He had been running with the underground for too long. Even though there were a lot of people like him in the underground, it was a closed society that was not easy to break into. And closed societies that are occupied by like-minded individuals breed ignorance. The members of these societies feed off of each other's thoughts, and little or no growth takes place. To make matters worse, the growth of IRC had turned the hacker scene into a big ego trip. Hacking was no longer about being good at hacking; it was about *telling* people that you were good. It was a PR campaign with no substance behind it. And when other hackers didn't believe you, you didn't prove it; you just typed in bigger letters and added a few curse words. A sign now hung on the door to the underground that said "no skills required; c'mon in." This time, however, Aaron decided to keep walking.

The stealing and the vandalism had never been about stealing and vandalism anyway, at least not in Aaron's mind. Well, maybe it was wrong in Aaron's mind, but it certainly wasn't wrong in the mind of his alter ego, Noid. To Noid, it was always about testing the limits of his curiosity, boredom, and a desire to see what he could get away with.

But Noid eventually came to the difficult conclusion that piracy often leads to more serious hacking, the type of hacking that Aaron knew in his heart he didn't want to get involved in. Piracy was like what people always said about marijuana: it usually led to something worse.

Aaron also noticed something else beginning to take place in the hacker underground that really turned him off to the whole scene. Script kiddies were beginning to pop up everywhere. IRC had become a battleground for egos, and nothing else. The BBS and the finesse of hacking and cracking had given way to a new scene that was no longer about earning respect and sharing information. Suddenly, the hacker scene was loud and obnoxious, and its young practitioners were into destroying data and preventing information from being shared. That wasn't why Aaron got involved in hacking. He no longer recognized the underground he once knew.

^ () < [] > () *^*

From that point on, Noid ceased to exist online. Aaron allowed his alter ego to ride off into the digital sunset. Of course, it wasn't quite that simple. He had good reasons for leaving hacking and cracking behind.

Aaron had been a mischievous kid. And that was putting it nicely. Getting dragged down to the police station, threatened with juvenile detention, and lectured by the cops about how lucky he was that he was a minor had an affect on his psyche, even if he didn't know it at the time. If he had been a few years older, the police said, he'd be keeping a mattress warm in a jail cell somewhere and doing God knows what else. This had a profound impact on Aaron as he started to look back on his hacking and cracking exploits. By this time, he had discovered his talent for hacking. Sure, some closed-minded "experts" were fond of classifying what he did as cracking, or software pirating, and not hacking. But Aaron knew better; he wasn't on the outside looking in. To him, cracking software was hacking without the liability of being traced on a network. "It was all the same. In the end, it was about getting in," he recalled. And he soon realized that he wanted a career in computer security. Suddenly, he had something to lose. Hacking or cracking—whatever you wanted to call it—had become a liability.

"By age 17, trying to see what I could get away with wasn't worth it to me anymore," he recalled years later. "The promise of the future was too great to get caught for something so stupid. There was no gain to what I was doing compared to what a career could give me."

The other factor that led Aaron out of the hacking community was the mass exodus from the Macintosh community that had taken place in the mid 1990s. The times had changed. The Mac hacker and cracker community had been overshadowed by the influx of Linux, Unix, and PC users. Aaron even sold his Mac and switched to a PC during his first year in college. But this made him nobody: the new kid on the block and the lowest hacker on the totem poll as far as the PC scene went. His ties to the only hacker community he had ever known had been severed for good.

At one point, he decided to check out the PC hacker scene as it existed on the various IRC channels. Actually, it was more than that. You might say that Aaron came close to falling off the wagon. His appearances on IRC were really a renewed attempt to get involved in the underground. But what he found when he got there convinced him that the decision to get out for good was the right one. "Everybody had huge attitudes and egos," he said, in an interview from his office at a relatively high-paying job in the private sector. "I knew they were all 12-year-old kids. But I was at the age where I wasn't really willing to kiss anybody's ass anymore. The whole IRC scene turned me off quickly. The Mac community was smaller, and the hackers were more willing to help each other. On the PC side, there were a lot people, and everybody was a prick."

^ ( ) < [ ] > ( ) *^*

Joe Magee never really left the hacking scene; he simply changed the definition of what a hacker is. Rather than allowing the thrill of the hack to lure him into a life of crime and possibly a new home in federal prison, Joe went to college, where he studied computer science. His hunger never subsided.

Joe was surprised to find that the same group dynamic that existed among hackers in the underground existed among his computer science classmates in college. He couldn't be friends or sociable with everybody. There were students whom Joe could tell knew their stuff, and others who didn't. The more talented hackers tended to talk to each other. It was as if college-level computer science classes had become the repositories for all of the underground hackers who had a conscience.

Joe ran into a few minor legal challenges while studying and "going crazy" on the network at Drexel University. He was, after all, a former member of a world where security was an afterthought. Now he was a legitimate user on what can only be described as a hacker playground—a university network. But his network roaming at Drexel was limited to wide-eyed, unadulterated discovery. That's what saved him. Universities have a hard time punishing students for taking part in anything that can be described as academic discovery.

Ice hockey also remained a part of Joe's life in college. But his player statistics were not what you would expect from a computer geek. With 39 penalty minutes and two game misconducts during seven games played, Joe acknowledges that he "was more of a goon" than a finesse player.

Today, Joe is the chief security officer at a major computer security firm. And what he sees taking place in the hacking world doesn't impress him. "If you're not hacking for scientific and educational reasons, then you're doing it for money," Joe says. "Why else would you do it? Credit card data is cash. And as far as Web site defacements are concerned, how does that help anybody? You're not going to get a job if you hack into a site."

Joe has also taken a firm, unapologetic stand against the argument that hackers who release what are known as "0-day exploits" (code that exploits known vulnerabilities in commercial software and is released into the wild on the Internet) are actually helping to improve Internet security. "There are proper ways to do that," he says. "Maybe you write exploit code, but you don't spread the word to everybody," Joe says, referring to some of his former hacker colleagues who are now employed by security firms and who have been accused of doing exactly that. "The public service argument is the biggest bullshit in the world."

Aaron's hunger also led him to college and a well-paying job in the computer security industry. Like Joe Magee, Aaron looks around the scene today and is not impressed with what he sees. Today, the wannabes make it nearly impossible for anybody to earn respect as a hacker, he says. Hacking, in his day, used to be difficult. There was a challenge to it, and hackers could take pride in breaking through tight code. Today, it's about destruction. And destruction has always been easy.

However, Aaron does see a benefit in the chaos that hackers are creating on the Internet. The paranoia that hackers are causing in the business world is a good thing, he says. "If hackers are sending that message, so be it," says Aaron. "They might be hackers, but they're also consumers. Hackers don't want their credit cards and identities to be stolen either."

3

The Hunt for Mafiaboy:
Operation Claymore

Shortly after 12 o'clock on Tuesday, June 8, 1999, students at Sisters High School in the small town of Sisters, Oregon, ran down the hall looking for Jon Renner. They found him in a classroom teaching a social studies class.

"One of the servers crashed," a student said, peeking his head through the door to the classroom. "None of us can get to our files or our personal Web pages."

Renner, who also served as the school's technology coordinator, wasn't particularly concerned by the news of the crash. The system had gone down before, and it was usually just a matter of making minor tweaks to restore operations. But there was something in the sounds of the kids' voices this time, a look on one of their faces maybe, that told him he should go have a look right away. After all, the server they were talking about wasn't your typical high school network server. This one powered a legitimate business enterprise.

It all started five years earlier. Renner, along with the assistance of a $50,000 grant from a local businessman, had helped the school set up a student-run Internet service provider (ISP) network. The ISP was called Outlawnet, Inc., after the Sisters High School nickname, the Outlaws. It was a small operation, designed to help pay for Internet access for the school district's 500 students. But the ISP had grown to the point where it was now serving more than 1,000 local residents and business customers in the towns of Sisters, Black Butte, and Camp Sherman. A group of 22 students, hand selected by a panel of teachers for their computer skills, helped run the company, developing Web pages, installing software for clients, and managing accounts. Each year a new Outlawnet class was selected, providing dozens of students with valuable real-world experience working in the computer industry. There was much to be proud of.

But on this day, less than a week after the school's fifty-first commencement ceremony, that sense of pride and hope for the future was replaced with a feeling of apprehension and fear. It took only a few minutes of inspection for Renner and another technician to realize that what had happened to their server was no glitch. There was nothing routine about what they had found.

A computer hacker had gnawed his way into the Outlawnet server. A vulnerable password had allowed the intruder to establish a shell account and inject himself into the network as a legitimate user. Nobody saw it coming. In fact, a hacker attack was unthinkable. Outlawnet wasn't Yahoo! or America Online; it was one of the small fish, a minnow in an ocean of whales. It was an ISP started on a shoestring and managed by well-meaning kids. What value could an attacker possibly see in hacking into Outlawnet?

The answer to that question would have to wait. The main Unix server had been obliterated and was inaccessible, even to the administrators. The maintenance programs that were reserved for use by the technicians were gone. More than 3,000 files, all of them belonging to the high school, had also been deleted. Dozens of user accounts had vanished as well. The intruder had also installed a sniffer program designed to capture insecure passwords and a mail relay system, effectively turning Outlawnet into a free e-mail relay station. It wasn't long before the telephone calls began pouring in from anxious customers who were worried about the impact of the virtual blackout on their businesses. This was a serious incident that required an immediate phone call to the local police.

The case was quickly passed to the Portland field office of the FBI. The Bureau's response was instantaneous and would leave Renner feeling nothing but the highest regard for the professionalism of the agents who were sent to investigate. Outlawnet may have been just a small-town ISP, but as far as the FBI was concerned, this was a crime with far-reaching implications, possibly even international implications. Launching a denial-of-service attack was a felony that could land you in prison, regardless of the size or economic status of your target. The FBI took every such attack seriously, and the attack on Outlawnet would be no exception.

On June 14, federal agents informed Renner that they planned to open an investigation into the attack, and that they had every intention of catching the hacker or hackers responsible. Renner pledged his full cooperation and vowed to pursue the hackers using every legal means at his disposal. In fact, he told the agents that if the intruder was based in the U.S., he planned to send Outlawnet's lawyers on the offensive to recover damages. Struggling ISPs like Outlawnet couldn't afford to allow attacks like this one to go unpunished. Renner was thinking about the need to restore not only his customers' confidence, but his students' as well.

It would be a week before Outlawnet technicians could restore all of the files that had disappeared and a full three months before all of the repairs to the system could be completed. Fortunately, Outlawnet had maintained a backup copy of all of the student files that had been deleted. But the attack would be costly. Software repairs and lost revenue from downtime were estimated to have cost the young company more than $11,000—money that would have gone to the school district to help pay for student Internet access.

Meanwhile, the FBI began to make progress in the investigation. They had tracked down a suspect in the U.S. by examining the system logs provided by Renner. However, the person fingered as the original suspect in the case turned out to be a legitimate business owner whose systems had been compromised and used as part of the attack on Outlawnet. The trail had temporarily run cold. But there was hope.

After answering a barrage of questions, the businessman handed the FBI agents a system log file containing an Internet protocol (IP) address. IP addresses are a series of numbers that act like street addresses for computers on the Internet. In this case, the IP address allegedly belonged to the computer that had first infiltrated the local businessman's system and then proceeded to attack Outlawnet. Slowly, the agents were getting somewhere. Although it was possible for hackers to fool another computer into thinking that a message came from an authorized IP address—a tactic known as spoofing—the FBI agents knew that if they acted fast enough, they would eventually find a link that would lead them to the real culprit. In this case, the first good lead was pointing to Sprint Canada.

^ ( ) < [ ] > ( ) *^*

Mark Gosselin had been with the computer crime squad of the Royal Canadian Mounted Police (RCMP) in Montreal for about three years when the FBI called and told him that they had traced a hacker incident in the U.S. back to an Internet account in Canada. According to the FBI, the hacker had taken down an ISP in Oregon using a high-speed digital subscriber line (DSL) account in Ohio that they were able to trace across the border to Gosselin's neck of the woods. It was December 1999.

At first glance, this seemed like a slam-dunk case. Gosselin was a 20-year veteran investigator for the RCMP. He had spent four of those years as a SWAT team member and the rest of the time doing old-fashioned detective work and handling counter-drug operations, fraud investigations, and criminal intelligence analysis. If the FBI had account information, it was just a matter of time before he would be able to trace it back to the source. Then he'd take a colleague along for a ride to arrest the "perp." Or so he thought.

Wherever the trail might lead him, Gosselin knew that he had a good case on his hands. When it comes to computer crime, Canadian law is just as tough on hackers as the legal system in the U.S. For starters, unauthorized use of a computer can land a hacker in jail for up to 10 years. In addition, destroying and altering data, known in Canadian law as "mischief to data," and obtaining passwords to fraudulently gain access to a computer also carry stiff 10-year sentences.

The first step in Gosselin's playbook was to obtain a search warrant for Sprint Canada. With the help of Sprint, Gosselin was able to uncover several e-mail aliases that belonged to an account at Delphi Supernet, an ISP in the Montreal area. But the account had been terminated a year earlier due to suspicion of hacking, a violation of the ISP's acceptable-use policy. Now things were getting interesting. Gosselin was beginning to make progress, but he still didn't have the smoking gun he was looking for. Even with account information, there was no way to tell for sure who was sitting in front of the computer at the time of the Outlawnet attack. And moving in too fast could blow any future case he might be able to make against the hacker, whom he presumed was a minor based on his past experience. He had to be sure. After all, there were literally tens of thousands of teenage boys living in the Montreal area who probably had the skill to conduct such an attack. And this wasn't the type of crime where you could conduct a lineup and have the victim ID the culprit. This was a faceless crime. And the evidence was thin. For the time being, Gosselin didn't have the proof that would enable him to get what he really needed, which was a wiretap.

Gosselin had narrowed down the source of the attack to a two-story, sea-foam-green and brick mansion-like house on Rue de Golf Street, nestled in an upscale neighborhood in the West Island section of Montreal, only 30 miles from the U.S. border. It was an area adjoining the chic St. Raphael golf course, set between the picturesque Lake of Two Mountains and the mighty St. Lawrence River. By most people's standards, it was a dream house in a posh setting. It came fully equipped with a two-car garage and a paved basketball court where the kids could play. And it was only 12 minutes by car to the nearest high school.

For the most part, the occupants of the house were like every other resident of the neighborhood, with a few minor exceptions. The owner, John Calce, was the president of a transportation company and was on his second marriage. According to neighbors, he was a husky, brash, unrefined loudmouth who liked to sit in front of his house in a sweatsuit yelling and cursing into a cell phone. He didn't pay much attention to his three sons, two of whom were brothers and the other a stepbrother from Calce's second marriage. The oldest brother was 17 and an aspiring actor who had actually managed to land an acting job in a local Montreal television show. Little was known about the stepbrother. But the youngest boy loved to play basketball. In fact, when he wasn't playing on the court at his house, he could be found playing guard for a local kids team called the Brookwood Jazz. When he wasn't in the mood for basketball, he helped neighbors and friends wash their cars. To many who knew him, there was nothing odd about him. He was a normal kid.

The young aspiring basketball player also loved computers. In 1998, when the two Delphi Supernet accounts linked to his residence were shut down due to suspicion of hacking, the young boy was only 12 years old. Gosselin would later suspect, but would never be able to prove, that the boy had likely learned about hacking from one of his older brothers. The kid probably wasn't sure exactly what his older brother was doing, but he knew in his heart that he wanted to do the same thing: hack into computers. To this day, authorities don't know exactly who was responsible for the hacking incidents that led to the cancellation of the Delphi Supernet accounts. The only thing that Gosselin could be sure of was that the house on Rue de Golf Street had a history of hacking.

Nobody could have predicted it, but the skinny, dark-haired, 14-year-old boy who liked basketball and girls would soon capture the attention of the entire online world and even the highest levels of the U.S. government. Unknown to Gosselin, the call from the FBI about a hacker incident at a relatively obscure ISP in Oregon was the beginning of what would become known as Operation Claymore. The world, however, would come to know Operation Claymore as the hunt for Mafiaboy, the most notorious teenage hacker since Kevin Mitnick.

^ ( ) < [ ] > ( ) *^*

Exactly one month prior to the attack on Outlawnet, a CIA officer tried desperately to warn intelligence officials in Europe that the Yugoslavian military facility they had targeted was, in fact, located one block away from the location where NATO pilots were about to drop their bombs. Unknown to the pilots or to the planners who had provided them with the coordinates, the facility, which was being used by the Serbs to support their brutal campaign of murder and torture of innocent men, women, and children in Kosovo, had been moved years ago. But by the time the CIA officer's concerns could be registered and acted upon by military officials in Europe, planes taking part in NATO's Operation Allied Force already were flying toward the target. When the smoke cleared the next morning, NATO officials awoke to the harsh reality that they had just bombed the Chinese Embassy, killing three people and wounding many others.

That was the day the cyberwar against NATO and the U.S. government started. It also was the beginning of the biggest U.S. government undercover operation ever to penetrate the hacker community.

The bombing of Serb military forces in Kosovo, along with the accidental bombing of the Chinese Embassy, had aligned hackers all over the world against the U.S.-led NATO alliance. Although Serb hackers had originally launched a defensive-information campaign designed to counter NATO's perceived dominance of the international news media, that campaign had now taken on an offensive character, targeting NATO and U.S. government Web servers. The objective was to use the skills of sympathetic hackers to disrupt NATO's ability to communicate news of the war effort across the Internet, which was increasingly becoming a primary source of information for millions of people.

Hackers from the U.S., Serbia, China, and Russia, all of whom either sympathized with or directly supported the Serbian cause in Kosovo, launched what was known as a ping-of-death attack against NATO Web servers located in the military alliance's headquarters in Brussels, Belgium. Ping is an abbreviation for Packet Internet Groper and refers to a method of determining whether a system is present on a network and operating properly. When an Internet user pings a server, he or she is sending a packet of information to the server and waiting for that server to send a packet back. It's sort of like making a telephone call just to ensure that somebody is home to answer the call. However, if you send enough pings to the server in a short period of time, you can overwhelm its ability to respond, especially to legitimate users who are trying to download information. That's exactly what happened to NATO's server. For several days the military alliance's message to the world about what was happening in Kosovo did not get out.

It was at this time that Bill Swallow, an experienced investigator with the U.S. Air Force's Office of Special Investigations who had been assigned to the FBI's Computer Intrusion Squad in Los Angeles, developed an informant buried deep within the Serbian hacker community. It would prove to be one of the most important contacts he would ever develop in his career.

As the war in Kosovo intensified, so did the hacker attacks against NATO, and especially against U.S. government systems. The FBI was now being told by the Defense Department that there had been several "clumsy" attempts to breach the network security perimeter of the Pentagon and the White House. It was clear from the forensic evidence that there was an overseas connection to the attacks, and that likely meant some sort of connection to Serbia and the U.S. war effort.

Calls were made from FBI headquarters in Washington, D.C., to FBI field offices all across the country instructing agents to make the increase in cyber attacks against government systems a top priority. As it turned out, one of the most important calls would be made to Charles Neal, the head of the FBI's Los Angeles field office, who years before had managed the well-known Kevin Mitnick hacker case. He was one of the FBI's most experienced cybercops and had actually helped develop many of the Bureau's computer-crime investigative techniques. Prior to joining the FBI, Neal had taught computer security at the college level and had also managed cyber security for companies in the banking and health-care industries. He knew what he was doing.

Shortly after receiving the call from Washington, Neal summoned Swallow to his office and asked him about the trustworthiness of the Serbian source he had been developing. It turned out that there were actually two sources feeding critical information to the FBI. One was a U.S.-based hacker who had close ties to the Serbian hacker underground, as well as family members still in Serbia. The other source was located in Kosovo and was a former war hero who had been working with the Serbian military in its anti-NATO information warfare campaign. Neal knew immediately that if these sources turned out to be who they said they were, the L.A. field office would have the best hacker intelligence capability of any office in the Bureau.

In 1999, only a handful of FBI agents in the Bureau had the technical capabilities and experience working with computers that Jill Knesek had. Assigned to the Los Angeles field office in January 1998, Knesek had been the agent who sifted through the mountains of forensic evidence collected against Kevin Mitnick. It was Knesek's ability to make sense of the data in the Mitnick case that helped send the notorious criminal cyberpunk to federal prison. Knesek brought to the FBI 10 years of computer security experience, including a stint as a computer specialist for the Naval Satellite Operations Center, where she was a programmer responsible for the maintenance of 15 navigation satellites. She knew her way around various types of computers and operating systems. Likewise, writing and deciphering hacker scripts came as easy to Knesek as signing her name on a sheet of paper.

As far as Swallow was concerned, there was no question that Knesek should accompany him on his trip to interview the U.S.-based hacker source. He was relatively new to the FBI team, and Knesek had all of the requisite technical skills. This assignment was too important to let pride or the desire to single-handedly bust a high-level case get in the way. That wouldn't have been a problem for either Swallow or Knesek anyway. They were consummate professionals who believed in defending the Internet from vandals and criminals as much as hackers believed in the inherent freedom of information and the legitimacy of hacking.

Swallow and Knesek met with the U.S.-based hacker, whose identity remains classified. The source immediately began feeding the FBI information on hackers located all over the world, including many in the U.S. Through this intermediary, Swallow and Knesek made contact with the Serbian hacker as well. Information began to flow from the front lines in Kosovo to Swallow and Knesek, and eventually to FBI headquarters in Washington. FBI headquarters then corroborated the information through other sources in and around Kosovo. It was a highly complicated information cycle that could easily be interrupted by the inherent lack of trust in the hacker underground and the difficult task of verifying the identity and truthfulness of the hacker in Kosovo. In the online world, a person rarely is what he or she appears to be.

The bombing campaign became Knesek's primary tool for verifying the location of the Serbian hacker mole. By checking the time stamp on the source's messages and then corroborating his reports of bomb blasts through FBI headquarters, Knesek was able to pinpoint the location of the hacker. The FBI was still able to get information faster than the U.S. news media, even CNN, so when the source reported that bombs had fallen close to where he was, Knesek and Swallow substantiated those reports with the help of the Pentagon and the Justice Department. The hacker was able to report bombing missions that were not yet known to the media. Knesek knew then that they had developed a source in the right place, at the right time—and, more important, he wasn't lying to them.

^ ( ) < [ ] > ( ) *^*

The U.S.-led air war in Kosovo lasted 78 days. Meanwhile, the FBI operation had ensnared a second U.S.-based hacker who would be prosecuted and turned into yet another critical government informant in the hacker underground. The identities of these initial hacker informants remain cloaked in secrecy. Investigations and cases are still pending.

The operation was a resounding success. Nothing like it had ever been done before, at least not in a cyber sense. Sure, the FBI had years of experience infiltrating criminal groups and other underground organizations, such as the Mafia, drug cartels, and white supremacy factions, but they had never attempted to infiltrate a subculture of faceless individuals to whom lies and deceit are not just a matter of self-preservation but a way of life. The hacker underground was unlike any underground organization the FBI had ever come up against. Even the mob had proven to be easier to crack and turn against one another. Hackers are tight; regardless of experience level, hackers don't run at the first sign of trouble. Instead, they morph into somebody else or take other evasive maneuvers that make it more difficult to track their whereabouts on the Net. And it takes months of proving yourself before you are considered a card-carrying member of the underground.

However, the FBI had attempted similar operations in the past. The Bureau was known to have infiltrated the hacker underground through various hacker organizations and meetings, such as local chapters of the 2600 hacker groups and various hacker conferences such as the annual DefCon conference held in Las Vegas. There had been successes, but not many. FBI agents and other government officials stood out like sore thumbs. In fact, each year the organizers of the DefCon conference hold a competition called Spot the Fed. The object of the game is simple: "If you see some shady MIB (Men in Black) earphone, penny-loafer, sunglass-wearing, Clint Eastwood, to-live-and-die-in-L.A.-type lurking about, point him out." Winners get an "I spotted the Fed" T-shirt.

Pointing out the federal agents at DefCon had become so easy that it was now a game. But for FBI agents online, identifying and locating hackers with serious criminal intent wasn't always so easy. In fact, it could be utterly impossible, like finding a tick in a tar bucket. In the online world, you need to be proactive to infiltrate the underground in a meaningful way. And sitting in the #dc-stuff (DefCon Stuff) IRC channel watching a bunch of washed-up, wannabe-hackers talk about getting wasted or falling-down drunk doesn't make you a hacker. Neither does just talking about hacking. You have to do it.

By late 1999, the Kosovo undercover operation was winding down. The number of attacks against NATO and U.S. government Internet sites had leveled off. But in the six months they had been working undercover, Swallow, Knesek, and about 150 FBI agents around the country, as well as local law enforcement and military agencies, had collected volumes of information on dozens of hackers who were known to be involved in criminal activities. "We basically found that we had infiltrated the hacker community," recalled Swallow. Indeed they had. And Neal was a seasoned agent who understood the value of intelligence. Without hesitation, a proposal was sent to the FBI and the Justice Department to keep the operation going. Getting approval was easy. Everybody in the DOJ wanted in. Thus, the first official undercover operation designed to penetrate the U.S. hacker community was born. The operation was so secret that its code name has still not been released to the public.

Like Gosselin in Canada, none of the members of the FBI's Los Angeles office had any idea what awaited them or how important their undercover work would soon become. The informants and investigative skills they were acquiring would prove critical to the FBI's ability to nab the teenager who was about to bring some of the biggest companies on the Internet to their knees.

^ ( ) < [ ] > ( ) *^*

The operation hit full speed in January 2000. Swallow had moved from a management position to the undercover team of agents who posed as teenage hackers online. Knesek helped Neal coordinate the nationwide dragnet to ensure that everything they did was legal. That proved to be one of the biggest challenges, and at times an obstacle.

Fortunately for the 40-something-year-old Swallow, he never had to meet face to face with any of the hackers he targeted. They never would have bought the idea that an old-timer like him was a hacker. But he did have to establish his credentials and a reputation among the hackers before he could collect any information that might assist a criminal prosecution. That meant that he and the others would actually have to hack and deface Web sites. It was no different than a Mafia boss handing a new member of the "family" a pistol and telling him to go prove his loyalty by knocking off a rival crime boss, only in this case, the legal tolerance for what the FBI might do to bolster agents' fake identities was much higher. Hacking, after all, was not a violent crime. It wasn't murder. Web sites didn't bleed.

Working through the U.S. Attorney's office could be a challenge. Every federal prosecutor, regardless of how little he or she understood about cybercrime or the hacker culture, wanted to take credit for the operation. The Justice Department touted the special training provided to these prosecutors as a sign that the department was no longer fooling around when it came to prosecuting hackers. What it really did, according to the agents in the field, was create a morass of confusion and competition that at times hamstrung investigations. Field offices couldn't make a move without prior approval.

Despite the political tug-of-war that was taking place between Washington and the various U.S. Attorneys' offices around the country, Neal and his team managed to obtain authorization from the Justice Department to deface various government Web sites to assist the agents in their effort to penetrate the underground. For the agents, next to knowing what they were doing online, getting approval was the most critical part of the operation. And it had to be timely. "We basically had to do a hack, or else we would lose credibility," recalled Neal. "It's no different than proving yourself in a gang or in the Mafia because the better hackers form their own chat rooms that are by invitation only."

The team eventually defaced about a dozen government sites to prove their mettle to the other members of the underground. They even convinced a few private companies to volunteer their corporate Web sites for defacement. Swallow and the other undercover agents sent copies of their hacks to the administrators of the Web site Attrition.org, an online archive used by hackers to show off the fruits of their labor. Attrition is the favorite repository and main source of celebrity for the teenage script-kiddies who do the bulk of the Web site defacements. Attrition doesn't care, and never asks, who the actual source of a defacement is; the only requirement is that the target be a legitimate Web site and not one that a hacker has set up for the sole purpose of hacking—that happens a lot, too.

Things went smoothly thanks to the help of the two hackers Swallow and Knesek had managed to bust a few months earlier during the Kosovo operation. As time progressed, there were more hacker stings. Many involved naïve teenagers who had become caught up in the allure of the hacker underground. Others weren't so naive, but were easily "rolled" into FBI informants. The serious criminals, on the other hand, often had a hard time appreciating the fact that they were "jammed up." Hackers, however, especially teenage hackers, are not hard to pressure once they know they are busted. You don't have to call them on the telephone 30 times a day to harass them, or order them to meet you at the coffee shop unexpectedly so they can recount their minute-by-minute activities. Those tactics might be needed for drug dealers and other hardened criminals, but not here.

Hackers who knew they were jammed up quickly became the FBI's trainers and consultants. They showed Swallow how to act in various IRC chat rooms and how to respond to questions and challenges from other hackers. In addition, they were critical to getting the FBI into the invitation-only chat rooms, where the serious hackers hang out. Without question, the best source of intelligence about a hacker is another hacker. Hackers are trusted members of an underground community where perceptions are reality, and where people who sound like feds are treated like feds, regardless.

During one of Swallow's late-night shifts, which often lasted 10 to 12 hours, a hacker approached Swallow and told him he had stolen 400 credit card numbers and stored them on a server in Germany. "Go ahead and get them if you want them," the hacker told Swallow, showing him a small sample of names and numbers to prove he had actually stolen the data. As far as Swallow was concerned, this wasn't a hacker who was trying to be nice or feeling eager to share the wealth. This was a challenge. It was a challenge designed to see what Swallow was made of and, more important, whether or not he was a fed.

Swallow wanted to do whatever it took to win the hacker's confidence. The credit card bandit was obviously one of the big fish: a real criminal. Credit card numbers are as good as cash. A single credit card heist can ruin a company's future. Customers usually don't return after an incident like that, especially customers who have had their personal information and credit card numbers stolen. With credit card spending limits reaching $5,000, $10,000, and higher, this hacker could easily be responsible for a theft worth about $4 million. Swallow was eager to nail this guy.

It wouldn't be the first time, however, that Neal would have to reign in his team. "I had to tell him you can't do it," recalled Neal. "Now we're conducting an investigation in a foreign country, we're violating international treaties, and you could have an international incident on your hands." Although there were no investigation guidelines to go by, Neal had to err on the side of caution. His agents were making it up as they went along; they were breaking new ground. And while that wasn't the answer Swallow was hoping to receive, he was a professional and understood what it meant to be screwing around with stolen credit cards that were stored on a server in a foreign country.

In situations like this, knowing how to stall for time can be crucial, especially from a legal perspective. Swallow eventually won the trust of many teenage members of the hacker underground, who would hit him up on the spot with urgent requests to help conduct a coordinated hacking attack. But this was where things got sticky. It was impossible, or nearly impossible, for Swallow and the others to make a move without the permission of the U.S. Attorney's Office. "We wanted to push the envelope, but we also didn't want to do anything that would cause the operation to be shut down," recalled Knesek. Often they were able to stall long enough to get permission. Their systems would crash "unexpectedly," they would have to make an urgent run to the bathroom, or somebody would show up at the door and spend the rest of the night talking their ears off. They used any number of excuses to get enough time to get legal permission to conduct a hack.

But there were countless other times when permission was not needed. During the course of the operation, Neal's team discovered thousands of compromised Web sites on the Internet. Government organizations and major corporations that managed critical national security–related functions or functions that were critical to the economy were told about the security breaches as soon as possible. But Neal's team had neither the time nor the resources to tell all of the companies that their systems had been compromised. He could easily dedicate all of his agents' time to making telephone calls to the thousands of companies that had no idea they had been hacked, but that would effectively shut down his intelligence-gathering operation. "I had to make a decision that we would not bother telling all of the 'flower shops' and opt for informing the sensitive government sites," recalled Neal.

Another major challenge facing the FBI during the undercover operation was nailing down the physical location of the hackers. Cybercriminals rarely tell you more than you need to know about them. These are the hackers who are adept at not getting caught and are not interested in publicity. They are criminally minded, and many of them understand the financial implications of their skills. Some of these act as mentors to the teenage script-kiddies and often enter a system right behind a script-kiddie, unnoticed and without missing a keystroke. Most, according to Neal, are found to be gainfully employed in the computer security industry and are actively involved in writing what are known as 0 Day exploits: malicious programs that send companies all over the country scurrying for software patches and consulting expertise. Ironically, the companies that employ these hackers are often the first to release patches or analysis of the security threat.

Teenage script-kiddies, however, had taken over the general IRC scene where Swallow often posed as an IRC channel operator. As an IRC member with operator status, Swallow was able to manage who was allowed to remain in chat sessions and who got booted off the channel. The loudmouths and the posers were often the first to get booted. It was impossible to have a serious discussion about real hacking tactics and tools with a bunch of kids interrupting all the time with inane observations that were off-topic and usually full of self-promotion and meaningless vulgarity. At the same time, these were the knuckleheads who most often told the FBI where they were located. Still, script-kiddies could be a real pain in the ass for the FBI and professional hackers alike.

And Swallow was about to meet the biggest, 125-pound pain in the ass the Internet world had ever seen.

$$*^*()<[]>()*^*$$

The first attack started on a Monday morning. It was February 7, 2000. Yahoo!, one of the Web's biggest information portals and e-commerce sites, was completely caught by surprise. The initial flood of data packets overwhelmed one of Yahoo!'s main routers at speeds higher than 1 gigabit per second, the equivalent of more than 3.5 million average e-mail messages every minute. The router recovered, but then Yahoo! lost all routing from one of its own major ISPs. There had been other problems in the past with this ISP, and administrators spent the first hour trying to eliminate known glitches.

Nobody thought that the company was fast becoming the first victim of the biggest distributed denial-of-service attack ever to hit the Internet.

Eventually, Yahoo! had to block all traffic coming in from its upstream ISP. This allowed administrators to restore basic network routing. But the picture of what was actually happening remained murky. The only thing Yahoo! administrators could tell was that their system had crashed under the weight of excessive Internet Control Message Protocol (ICMP) traffic. Computer networks use ICMP messages to troubleshoot problems, such as a router that is unable to transmit data packets as fast as it receives them. ICMP messages communicate these problems between systems automatically. And that's when the Yahoo! administrators realized that the problems their network was experiencing were not the result of a random glitch. This was a deliberate attack.

Yahoo! technicians immediately began filtering out all ICMP traffic. But then all of the traffic that had been blocked by the original slowdown began to pour in and choke the Yahoo! servers. One of Yahoo!'s ISPs managed to capture some data on the attacks, and the Yahoo! technicians noticed that a large number of their peering circuits—the major national ISPs with which they share data—were unwittingly taking part in the attacks. In fact, one of the traces led Yahoo! technicians right back to one of their own computers; Yahoo! systems were even attacking other Yahoo! systems. This was a highly distributed attack that used many computers as pawns, better known as zombies, in the attack. And a highly sophisticated hacker or group of hackers was likely responsible, according to Yahoo! experts who were doing battle with the attackers. Who else could have been responsible for such a massive denial-of-service attack?

"It seemed the attacker(s) knew about our topology and planned this large scale attack in advance," wrote a Yahoo! system administrator a few days after the initial flood of packets took down the Yahoo! network. "It seems that this is definitely a DDoS attack, with attacker(s) [who are] smart and above your average script-kiddie. Attacker(s) probably know both Unix and networking pretty well."

There it was: the first detailed analysis to come out of the first company to be hit. It was clear that Yahoo! was dealing with a hacker who knew what he was doing and who took the time to learn about his target and plan the attack. There was no way that what Yahoo! administrators were witnessing was the work of a kid who wanted simply to find out whether the DDoS scripts he had downloaded from the Internet actually worked. This attack was the work of a pro, who probably had help—Yahoo! administrators were convinced of at least that much. By the time it was over, the Yahoo! attack alone would involve enough data to fill 630 pickup trucks with paper.

Later that night, Swallow poured himself a cup of coffee, sat down at his computer, and prepared for another long night of mostly meaningless chat sessions with mostly insignificant script-kiddies. The nights were getting long; none of the real hackers showed their virtual faces until the wee hours of the morning. But on this night, Swallow, acting as channel operator on one of the IRC channels frequented by hackers, noticed that somebody with the nickname Mafiaboy had popped up on the scene. Swallow had noticed him before, or at least somebody who had used the name Mafiaboy (there was no way to know for sure whether it was the same person). But any doubt about the true identity of this Mafiaboy was soon put to rest. The kid turned out to be true to form. This Mafiaboy was the same loudmouth script-kiddie with whom Swallow and the others had exchanged words in the past.

Tonight, Mafiaboy was bragging about his "skilz." The other hackers on the IRC channel quickly grew tired of Mafiaboy's ranting. The exchanges soon degenerated into outright name calling, relying on the usual collection of expletives. Cursing was one of Mafiaboy's more refined skilz. But this wasn't what Swallow had in mind for the rest of the evening. The members of the chat room grew so tired of Mafiaboy's bragging and bogus claims that he had pulled off a major hack that Swallow booted him out of the chat room.

Shortly after 9 A.M. on May 8, Buy.com, an online retail store, issued investors its initial public offering of stock. The future looked bright for this Internet business as it caught the wave of dot-com mania. However, at 10:50 A.M. Buy.com system administrators were battling a massive denial-of-service attack involving 800 megabits per second of incoming data, more than twice the Web site's normal load. The attack threatened to keep the retailer offline indefinitely. Later that afternoon, the Web's most popular auction site, eBay, also reported significant outages of service, as did venerable online book retailer Amazon.com. If there was a bright side to this situation for Buy.com executives, it was that their fledgling company was not alone. Nor was it the last of the victims.

When Swallow came on duty that evening, he was once again confronted with the brash young hacker who called himself Mafiaboy. By this time, Swallow was aware of the situation on the Internet and was hoping to find leads through the IRC. Mafiaboy once again claimed responsibility for the attacks. But there was no way that Swallow or the other hackers on IRC that night were about to fall for that. This was vintage Mafiaboy, they thought: always the loudmouth, annoying script-kiddie who everybody knew. That's when Mafiaboy put a challenge to the rest of the IRC members.

What do you want me to hit next, he asked the hackers on the board.

Swallow and the others ignored him. This guy ranked at the top of the "bogometer"—or bogus meter—they said. He's a real cretin, a loser with no real skilz. Then somebody suggested that CNN might be a good target, as would E-Trade.

Fine.

Within minutes, CNN's global online news operation, as well as 1,200 other Web sites that CNN hosted worldwide, started to grind to a crawl. By the following day, Datek and E-Trade, both online stock-trading companies, entered crisis mode as sporadic outages of Internet operations threatened the health of the financial markets. Slowly, as the scant forensic evidence on the source of the attacks began to be pieced together, it became clear that dozens of computers had been hijacked and used in the attacks. Vulnerable computers at the University of California in Santa Barbara, the University of Alberta in Canada, and in Atlanta and Massachusetts had been turned into zombies—as many as 75 computers around the world. The intruder had planted malicious software on these systems that had turned them into autonomous launching pads for denial-of-service attacks.

This was a true crisis, the one that all of the experts had been warning about for years. Fears began to circulate that the attacks were the beginning of what national security officials had been calling an electronic Pearl Harbor: a surprise cyber attack designed to cripple the U.S. Internet infrastructure. The Internet was in a full-fledged meltdown. The media latched onto the story like it was the end of the world. The economy would certainly enter a death spiral if the attacks continued. But would the attacks continue? How many other systems had been infected and turned into zombies, time bombs waiting to be detonated remotely by their hacker master? The answers to those questions were critical. Nothing less than the public's confidence in the future of the Internet economy was at stake.

The FBI needed to find the hacker who called himself Mafiaboy. And they needed to find him fast.

^ ( ) < [ ] > ( ) *^*

Knesek was in a hotel room in a rural part of Alabama, where she had been serving a search warrant against another hacker the FBI had ensnared in the undercover dragnet, when the phone rang. It was Neal.

"We've got a major problem on our hands," he said. "A hacker is hitting all of the major ISPs and e-commerce sites, from Yahoo! to Amazon to CNN." Initial evidence showed that the attacker used telnet through Wingate proxies to cloak his activities. Most of the compromised machines that had been turned into launching pads were located at universities and ran the Red Hat Linux 6.1 operating system, he told her.

Knesek immediately hopped on the Internet to try to find some leads. She, too, had worked undercover posing as a teenage hacker for a few months before taking over the legal coordination of the operation. But there was only so much she could accomplish from Alabama. Leads were still hard to come by. And by the end of the week, she was back in the L.A. office.

Neal had decided early on that the L.A. office would attack the investigation from an intelligence perspective. Other FBI field offices, such as the San Francisco office, had decided that they were going to approach it from a technical standpoint. But Neal knew he had the best sources of intelligence that the hacker underground could offer. That was, after all, the whole point of the undercover operation he had been running for the past year. Eventually, it would be a combination of the two approaches that would enable the FBI and the Canadian police to home in on the real Mafiaboy.

Finding the real Mafiaboy would be difficult. Within days of the first attacks, the false confessions started pouring in. Dozens of calls a day had to be fielded, and dozens more appeared on the Internet via IRC chat rooms. Everybody wanted to claim responsibility for the biggest hacker news story since the MOD boys crashed the long-distance telephone network in 1990. Just trying to filter out all of the bogus confessions was starting to turn relatively normal working hours into 80-hour weeks for most of the FBI team. Then there was the challenge of deciphering who the real Mafiaboy was from the three that the FBI knew of and the dozens of other hackers who suddenly pretended to be Mafiaboy online.

Information continued to pour in to the FBI from the victim companies. The major networking companies, along with Exodus Communications, Inc., which provided Internet services through its Los Angeles hosting facility for some of the big-name companies that had been hurt by the attacks, had started to crunch through router logs and were gradually beginning to piece together a picture of what hosts had talked to each other during the attacks. A portrait of the real hacker was emerging.

Neal dispatched several agents to the Exodus network operations center to examine some of the computers that were involved in the attacks. But when the agents arrived, the guards turned them away. The Exodus guards demanded that the agents present more than just FBI credentials before they would let them enter the facility. Neal was pissed. He immediately jumped on the phone and called Exodus. He started with the main number, but from there one employee handed him off to the next and to the next, and so on. Growing impatient with the delay and the critical time that was being lost in the investigation, Neal continued to escalate his request up the corporate chain of command at Exodus. Eventually, his call landed on the desk of Bill Hancock, Exodus's brand-new chief security officer. Hancock had been on the job literally one day when Neal's call came in. Old friends from years before, the two were surprised to run into each other under such strange circumstances. After a very abbreviated discussion to get reacquainted, Neal had the cooperation he needed from Exodus. The data collected from the systems at the Exodus facility would prove critical to the investigation.

On February 12, Dell Computer Corporation reported that its systems had been hit with a barrage of Internet traffic. Once again, Mafiaboy went online and renewed his public relations campaign, claiming responsibility for the Dell attack as well as all of the previous incidents. Several security experts in the private sector, as well as other hackers, captured the following chat session log and sent it to the FBI. The chat log recorded that night in the IRC hacker room #!tnt details a conversation between Mafiaboy, who had changed his nickname to anon (for anonymous) and several other hackers using the names T3, Mshadow, and swinger:

ANON: SNIFF ME FBI!!

ANON: t3, can u connect to dell? Some people say yes, some say no

T3: I can't browse the web

ANON: fools don't know what cache is. telnet to port 80

T3: period, my modem is totally [f***ed]. Everything times out.
Uh no thanks.

SWINGER: anon. It lods but like slowly

SWINGER: heh no its really lagged

MSHADOW: Are you just hitting it with a stream attack?

ANON: mshadow no, my personal attack

MSHADOW: hehe, what kinda packets?

ANON: which is spoofed ++

ANON: its sorta a mix. A new type and syn

T3: spoofed++, lol

T3: it's either spoofed or not spoofed, there is no "elite" spoof, but I think there's methods to trace spoofed packets, if you catch it while it's flooding

MSHADOW: yeah, they have go router to router to trace the packets back, only takes like 20 min

T3: mafiaboy, so who's next after dell

MSHADOW: You know wait till they talk about it on msnbc

SWINGER: ms should be next, and drop the chat server

ANON: t3 tonight I put this computer in the fireplace

SWINGER: heh

MSHADOW: haha

ANON: I aint joking

MSHADOW: why don't you just take out the hd and kill that then put a new one in

ANON: mshadow I don't want to take ANY chances

MSHADOW: really. and talking on IRC is not a chance?

ANON: what can irc prove?

T3: mafia

ANON: uhm [f*** it], [f***] the fire place, sledge hammer instead

T3: it's spoofed, they can't catch you. I need to get away from you before I get busted for being an accomplice or some shit.

ANON: t3 don't give a [f***]

MSHADOW: haha

T3: heh

ANON: don't take chances

T3: aren't you going to go out with a bang at least?

ANON: yes

MSHADOW: drop like 10 core routers :\

ANON: no

T3: what are you gonna do

ANON: Microsoft

ANON: Microsoft will be gone for a few weeks

T3: HAHAHAHAHAHAHAHAHAHAHAHAHAHA

T3: oh man, that's evil

ANON: MAYBE, I'm thinking something big, maybe www.nasa.gov, or www.whitehouse.gov, maybe I'm just bluffing

T3: I need to get away from you before I get busted for being an accomplice or some shit

ANON: t3 "hit the router"

ANON: the whole router list

ANON: I know what im doing

ANON: yahoo.com

T3: haha

T3: So Mafiaboy, it was really you that hit ALL those ones in the news? buy.com, etrade, eBay, all that shit?

ANON: you just pin em so hard they cant even redirect

ANON: t3 maybe. who knows. would only answer that under ssh2 [a secure, encrypted connection].

T3: haha

ANON: i might pmg the hd and sledge hammer and through it in a lake

T3: they say you're costing them millions

ANON: surprised I didn't get raided yet, t3, they are fools

Despite his admissions, FBI agents and seasoned members of the hacker community had a hard time believing that the kid who called himself Mafiaboy was capable of pulling off the attacks. He wasn't skilled enough, and the attacks were too sophisticated, too distributed for one minor-league script-kiddie to do on his own. In addition, private-sector security experts were telling the FBI that the February 7 attack against Yahoo! differed significantly from the attacks that took place later in the week, and that it was highly likely that multiple hackers were involved.

In retrospect, the FBI and the rest of the hacker community saw that they had missed the first clues. The Mafiaboy they were all familiar with was, in fact, the Mafiaboy responsible for the attacks. Although there's no evidence to suggest that had Swallow and the other hackers believed Mafiaboy when he first started to brag about the attacks they could have stopped them from happening, the initial clues were missed, the alarm bells failed to ring. The FBI's perceived ineptitude emboldened Mafiaboy. He boasted that he would never be caught. More important, his reference to putting his computer "on the fireplace" was not an idle threat. He would eventually throw his hard drives into a nearby lake. They would never be recovered.

For the next two days, Neal and his team of FBI experts scoured the Internet for clues to the real identity of the hacker known as Mafiaboy. On February 14, they found the following Web page:

www.dsupernet.net/~mafiaboy.

That was a Web site belonging to a user of Delphi Supernet in Canada. Shortly thereafter, forensic evidence came in that linked the Dell attack to an Internet account with an ISP in Montreal called TOTALNET. The FBI now had two pieces of evidence pointing to a Mafiaboy in Canada. This was critical evidence because there were at least two other Mafiaboys who had been online using that nickname often enough to force the FBI to search them out. The more suspicious of the two alternative Mafiaboys ended up being a college student in New York. But although the FBI agents in New York were convinced that they had their man, Neal was adamant that the real culprit was sitting across the border in the Montreal area.

A third piece of critical evidence bolstering Neal's contention that the Mafiaboy in Canada was the Mafiaboy whom authorities should be going after was the data from the initial attacks, which had been preserved by system administrators at UC Santa Barbara. In addition to a complete backup copy of log files documenting exactly what the hacker had done on their system, administrators at the university also produced a copy of the attack tool used. The attack software was registered to a user named Mafiaboy and another user named Short. Both handles were later proven to have belonged to the same person. However, more important to the eventual prosecution and conviction of the teenage script-kiddie was the warning that the tool's author had given to all the hackers who downloaded it:

```
WARNING: Using this program on public networks
is HIGHLY illegal and they WILL find you and
put you in jail. The author is no way responsi-
ble for your actions. Keep this one to your lo-
cal network!
```

The denial-of-service tools used by Mafiaboy included Stacheldraht, the German name for Barbed Wire, a variant of the dangerous Tribal Flood Network (TFN) software that swarms target systems and overloads them with data requests. However, like most script-kiddies, Mafiaboy did not write the tools or the code he used in the attacks. A more sophisticated hacker based in Germany, known as Randomizer, is believed to have been the author of the tool used by Mafiaboy. In addition, another 20-year-old German hacker who went by the name Mixter was the brains behind the original TFN attack software. The FBI quickly dispatched agents to track down and interview Mixter. Shortly after being interviewed, Mixter was eliminated as a suspect in the February attacks. Mixter later denounced the attacks as criminal and wrong.

^ ( ) < [ ] > ( ) *^*

The call came in from Washington on February 14. The FBI needed the help of the Royal Canadian Mounted Police to nab a hacker named Mafiaboy, whom they suspected was living somewhere in the Montreal area. The FBI and the RCMP had a long history of working together, and the RCMP immediately agreed to help. Thus marked the official beginning of Operation Claymore.

RCMP agent Gosselin had moved on to many other cases since his investigation of the hacker attack against the student-run ISP in Oregon. That case had gone nowhere and was now a distant memory. But with the call from the FBI, Gosselin was appointed the lead investigator charged with tracking down the real Mafiaboy. The choice of Gosselin would prove to be one of the RCMP's most important decisions in the case.

The next morning, February 15, Gosselin executed a search warrant for the systems at the Delphi Supernet and TOTALNET offices in Montreal. He discovered three e-mail accounts registered to a Mafiaboy:

mafiaboy@dsuper.net

mafiaboy@total.net

pirated_account@total.net

Although Gosselin had discovered accounts with the Internet handle Mafiaboy attached to them, this didn't mean that those accounts belonged to the guy Gosselin was looking for. One of the e-mail messages discovered had an Internet protocol address linked to it, but it turned out to be a hacked account that belonged to a real estate broker: pirated_account@total.net. This account would later be identified through phone tap and trace correlations to Mafiaboy's residence; he'd obtained the unsuspecting couple's account password and was dialing in and using the account from his house.

Once again, Gosselin started the tedious process of pouring through account information and cross-checking telephone numbers, credit card numbers, and names on accounts and mailing addresses. Everything was different, nothing matched up—except for one phone number. It was the phone number that most ISPs and credit card companies ask their customers to provide as an alternate contact number. For reasons unknown to him at the time, that number looked familiar to Gosselin. He did a search for addresses against that phone number, and the search returned a match: Rue de Golf. Now things were beginning to add up. Gosselin recognized that address.

It is at times like these that years of experience doing old-fashioned detective work come in handy. Despite the advance of technology and all the talk about how difficult it can be to find computer outlaws, agents like Gosselin know that it almost always comes down to hard-nosed detective work, pounding the pavement. Gosselin rifled through his old files looking for a lead. One of the first files he thumbed through was the one from the hacker incident at the Oregon ISP, Outlawnet. The address and telephone number matched the address and telephone number of the suspect in the Oregon ISP case. Mafiaboy was still active. This time, however, Gosselin felt like they had their man, or at least a solid address for the real Mafiaboy. "It looked good. It looked very good," he recalled. Bolstering his suspicions were a series of complaints that the ISP had collected over the years about the users of this account. It seemed that others had already fallen victim to a hacker who had been traced back to the Delphi Supernet account.

The only reason Gosselin had not busted Mafiaboy months ago was the lack of evidence; had he been able to prove probable cause, he would have put a wiretap on the house in December. It's hard to say what might have happened had Gosselin no longer been detailed to the Montreal office of the RCMP when the February DDoS attacks occurred. Another investigator might never have been able to make the connection. Critical time would have been lost in the investigation.

In Washington, D.C., President Bill Clinton summoned dozens of senior industry representatives and members of the national security community for an urgent meeting at the White House. The recent wave of denial-of-service attacks against some of the Web's biggest companies was a threat "to our whole way of life in America," the president told the gathering in the second-floor cabinet room. "I don't think it was a Pearl Harbor," he said. "We lost our Pacific Fleet at Pearl Harbor. I don't think the analogous loss was that great." The attacks were, however, "part of the price of the success of the Internet," Clinton said.

Clinton started the meeting with a request to have a few of the industry experts explain to him how the attacks happened and how such a devastating meltdown in Internet services could be possible. Rich Pethia, an expert with the Computer Emergency Response Team at Carnegie Mellon University, was the first to offer an explanation, followed by Tom Noonan, chief executive officer of Internet Security Systems, Inc., and Vint Cerf, a senior vice president at MCI Worldcom. Each expert ran down the technical reasons for the massive vulnerabilities that contributed in one way or another to the success of the attacks. The most intriguing explanation, however, was offered by Witt Diffie of Sun Microsystems. Sitting across from the president at the large, highly polished table that was normally reserved for high-level discussions of major foreign and domestic policy issues, Diffie told the president that the attacks were akin to a breakdown of democracy. It was as if the president had lost an election, not because people didn't vote for him, said Diffie, but because somebody stole votes and cast them in favor of his opponent. This was an analogy Clinton could understand. But as many of the experts would later say, Clinton didn't need any help understanding exactly what the attacks meant. No, they weren't an "electronic Pearl Harbor," but they were more important than a lot of people were letting on.

Clinton and some others present at the meeting had a more difficult time understanding the argument being made by Peiter "Mudge" Zatko, a hacker turned security consultant and a former member of the hacker group L0pht Heavy Industries. Many interpreted the presence of Mudge at the meeting as an example of the ultimate irony. The L0pht group was responsible for developing one of the most powerful password-cracking tools on the Internet, called L0phtCrack. A former member of a hacker group that government officials once denounced as part of the problem was now sitting in the White House advising the president on how to protect government systems. Seated between National Security Advisor Sandy Berger and Attorney General Janet Reno, Mudge's long hair and cyberpunk demeanor seemed out of sync with the stiff, neatly coiffed gathering of officialdom arrayed around him. To the surprise and dismay of some, he urged the government not to criminalize the act of creating offensive hacking tools. According to Mudge, if the government criminalized the act of developing such tools, the security community would be barred from discovering effective ways of defending against them.

Mudge's advice fell mainly on deaf ears—not surprising, since most executives and government officials are good at pretending they understand what hackers are all about, when in reality, most senior business executives and government officials just don't get it. They don't understand that security isn't a hacker thing. They have an even harder time wrapping their bureaucratic minds around the fact that hackers aren't about crime, and that the act of hacking isn't about breaking the law. Of course, many of Mudge's former colleagues in the hacker underground are partly responsible for those misperceptions. In many ways, hackers have only themselves to blame for what the media writes about them.

Although the president's advisors did not offer a single, coordinated view of the hacking and security world, Clinton demonstrated a firm understanding of the subject matter and the serious impact that the February attacks could have on the U.S. economy if they happened again. Clinton's concern about the situation was in stark contrast to that of the chief executive officers from industry who chose to send junior executives who did not have the authority to make policy decisions on behalf of their companies. Although Clinton invited the CEOs, most stayed away to avoid being put on the spot by Clinton, whom they feared would ask for commitments on new Internet security policies. It was a signal that the private sector was not really concerned about the attacks. Experts would later estimate that the attacks cost businesses more than $1.7 billion in repairs and lost revenues.

And now there was mounting evidence that a skinny, defiant, and mischievous 14-year-old kid in Montreal was to blame. Go figure. In retrospect, the fact that a 14-year-old had managed to take down the biggest companies on the Internet is no surprise. Packet-flooding techniques are widely known and understood throughout the hacker community, and the tools to conduct such attacks are available free on various hacker Web sites. In the past, hackers had to infiltrate each machine separately and launch individual versions of the denial-of-service tool. Now, however, automated scripts make it possible for unskilled teenagers to conduct massive scans for vulnerable computers, install the DDoS software, and then order those computers to start flooding other systems at will.

When it comes to DDoS attacks, bandwidth—the size of the communications pipe—can be as important as the number of hosts involved in the attack. High-speed networks, like the 25 university networks that Mafiaboy compromised, are highly effective DDoS-attack launching pads. University systems provide fertile ground for malicious hackers because of their lack of security. In the end, however, the success of a DDoS attack often comes down to the number of systems that are allowed to operate on the Internet with known vulnerabilities in them—and that includes government and corporate systems as well. Had the computers that Mafiaboy used in the attacks been secured properly by the system administrators who were entrusted with doing so, he may never have been able to carry out his week-long hacking spree.

And there had been plenty of warnings. In the fall of 1999, the FBI's National Infrastructure Protection Center began receiving reports about the existence of new software tools capable of launching massive distributed denial-of-service attacks on the Internet. The tools were the same ones that Mafiaboy would later download and use in his attacks. Because of its concern about this new threat, the NIPC issued warnings to government agencies, private companies, and the public in December 1999. Nobody listened.

At the same time that the NIPC had issued its warning, the agency had developed a tool that network administrators could use to detect the presence of DDoS software agents on their systems. It was the only detection software of its kind in existence at that time. Therefore, the NIPC took the unusual step of releasing the tool to the public in an effort to reduce the level of the threat. It posted the first variant of its software on the NIPC Web site on December 30, 1999. A press release was issued, as were three updated versions of the software that not only fixed earlier deficiencies but also enabled the tool to run on different operating systems.

The FBI and its cybercrime prevention arm, the NIPC, had done the job everybody expected of them. But unless companies and universities actually downloaded and used the software to check for the presence of malicious software on their systems, no amount of warning would be enough to prevent an attack.

By February 16, word of Gosselin's success in tracking down a solid lead was passed to the FBI. Plans were being made to obtain legal authority to install dialed-number recorders (DNRs), commonly known as pen registers, on the telephone lines leading into and out of Mafiaboy's residence on Rue de Golf street. A DNR is the equivalent of a caller ID system that tracks all outgoing calls made from a suspect's telephone to show that the suspect is communicating with known criminals, or in this case, with known ISPs. DNRs are a critical tool investigators use to locate accomplices and, if necessary, to demonstrate the need for full wiretap authority. DNRs were used extensively during the investigation into the Masters of Deception in 1990 and helped law enforcement officials uncover a large group of cyber trespassers who enjoyed crashing the AT&T long-distance network.

The DNRs on Mafiaboy's telephones were in place on February 18, the day that the FBI's Jill Knesek arrived in Montreal. But DNRs have their limitations. You can't capture voices with DNRs, only phone numbers and dates and times of calls. But the RCMP's tactics were about to change—and with that change, investigators would learn details about Mafiaboy and his family that would not only seal the young hacker's fate but also would change the way a lot of people thought about teenage hackers.

^ ( ) < [ ] > ( ) *^*

When Knesek arrived in Montreal, she immediately took on the role of intermediary between the FBI's Washington headquarters, the U.S. Justice Department, and the RCMP's lead investigator on the case, Gosselin. The number of requests for information coming from Washington was mind boggling, and the RCMP had grown increasingly distrustful of the officials in Washington. They became reluctant to share a lot of information with them. After all, the case now fell clearly within Canadian jurisdiction. The RCMP was the lead agency responsible for what would be the first data interception operation of its kind for Canadian law enforcement.

Knesek remembers being very impressed with the amount of information the RCMP had already managed to collect on Mafiaboy by the time she arrived in Canada. She was also surprised when she met Gosselin for the first time. Although the name hadn't originally rung a bell for her, the two law enforcement agents realized shortly after meeting that they had recently attended specialized computer crime and hacking investigation training together at an FBI training session in Baltimore. During the next several weeks, Knesek and Gosselin would build "a very good rapport," and one that would be critical to the flow of the investigation.

Within four days of the setup of the DNRs, investigators discovered another TOTALNET account registered to Mafiaboy. This time, however, the account belonged to the transportation company owned and operated by Mafiaboy's father. Despite the cancellation of the previous accounts two years earlier, it was now obvious that Mafiaboy had multiple ways of connecting to the Internet and identifying himself to others. There were hacked accounts, legitimate accounts, and accounts that ostensibly belonged to family members. Although the RCMP had narrowed down the search to a single residence, a major challenge still lay ahead. Who was sitting in front of the computers during the attacks? Again, Gosselin and the FBI were confronted with a dilemma: move in too soon, and the case would collapse. Mafiaboy, whoever he was, would go free.

With five people living in the suspect's house, including three teenagers, there was only one way to find out the true identity of Mafiaboy and whether he had any accomplices in the attacks. On February 25, the FBI and the RCMP obtained a court order to intercept all private communications of Mafiaboy and his entire immediate family. That meant a full-blown wiretap and a massive data collection operation focusing on all telephone conversations and computer and Internet activity that took place in the house. They would have 60 days to collect all the evidence they needed before they would have to reapply for the court order.

But at this point in the investigation, something else happened that raises questions about how Mafiaboy may have been caught. Although the details remain a closely guarded secret, the FBI and the RCMP acknowledge that an informant played an important role in identifying Mafiaboy and letting agents know when he was online. There are still questions about who the source might have been and how the source might have verified that Mafiaboy was online. The RCMP and the FBI are not talking. However, RCMP investigators have acknowledged that at this point in the investigation, they had determined that one of Mafiaboy's brothers had used the nickname Mafiaboy in the past, a fact that would eventually "help the investigation," according to RCMP Staff Sergeant Robert Currie, who headed up the Computer Investigative Support Unit. Was the informant one of Mafiaboy's local friends, who were also known to be involved in hacking, or was it somebody living in his own house? We may never know. Mafiaboy may never know. Regardless, the RCMP and the FBI already had a good idea about the person they were trying to build a case against in that fancy house on Rue de Golf Street when the wiretap was authorized.

Data interception operations began on February 27. TOTALNET created a preset range of IP addresses to be used only for Mafiaboy's suspected accounts, enabling investigators to focus closely on his activity. Data interception servers were set up at the ISP as well. The information began pouring in immediately. Each day's capture was reconstructed using proprietary software developed by the FBI. The job of collecting, managing, and analyzing the deluge of information fell to Currie.

As the head of the RCMP's Computer Investigative Support Unit, Currie actively monitored all Internet activity originating from Mafiaboy's residence and sifted through it for clues that would help investigators build a case against the teenager. As Currie would soon find out, capturing the data is the easy part. The tough part comes in separating different activities, such as Web surfing, online gaming, and e-mail, and then trying to decipher with whom Mafiaboy might have been communicating.

Mafiaboy had his active days and his not-so-active days, but on his active days, the teenager often operated until three or four o'clock in the morning. Currie set up his system to conduct the daily download of raw data intercepts shortly after four o'clock, when Mafiaboy was known to quit for the night. When the operation ended 43 days later, Currie had collected 7.6 gigabytes of raw data.

Most of Mafiaboy's online activity involved Web surfing, online gaming, and boisterous IRC chat sessions. But he was set up pretty well, making good use of a dual-boot Unix and Windows NT system. He downloaded Back Orifice scanner, a Trojan horse (back door) program developed by the infamous hacker group Cult of the Dead Cow (cDc). The inexperienced teen hacker also worked feverishly trying to figure out Netcat, a port listener that gives a hacker command-line access to a system. But Mafiaboy struggled with Netcat, demonstrating for officials that he wasn't that good after all. If a hacker is any good, he should have Netcat figured out relatively early in his career. But Mafiaboy was learning, and he had an aptitude for hacking.

During one telnet session, agents watched Mafiaboy in real time as he attempted hacks and had to retype commands three, four, or five times before he got them right. In addition, he always seemed to be accessing accounts using logins and passwords that other hackers had given to him. One hacker sent him a file containing account information for 20 different university systems.

When he wasn't racking his brain trying to decipher Netcat or busy retyping basic hacking commands, Mafiaboy was setting up rogue file transfer protocol (FTP) sites on the systems he had compromised as part of the February DDoS attacks. He used the FTP sites to trade Sony Playstation CDs and DragonBallZ Japanese animation videos. He downloaded Unix scanners to probe for compromised Internet accounts. And he hung out on the systems of any of 25 universities around the U.S. and Canada.

In addition to being sort of a street urchin with street-smart friends who also were into hacking, Mafiaboy had a lot of friends on the Net. IRC was a central part of his Internet existence. In addition to frequenting the #exceed, #shells, and #carding IRC chat rooms, Mafiaboy was part of an organized group of hackers called TNT, as evidenced by his IRC conversations in that invitation-only chat room. Eventually, he left that group and said that once the dust settled from his TNT experience, he would start his own hacker group and take down all kinds of sites on the Internet. He enjoyed the fame he received from hacking. But there was one problem facing the young hacker: he wasn't as good as the software he used made him appear to be.

⅄ ^* ( ) < [ ] > ( ) *^ *

In March, Mafiaboy's father installed a digital subscriber line from Sympatico-Lycos, Inc., one of Canada's major ISP and Web hosting companies. On March 16, data interception operations on the Sympatico DSL modem started.

There was so much data to capture that Currie set up a mini-lab in the basement of his home so that he could conduct downloads in a more timely manner, as well as watch his kids from time to time from his RCMP office through a digital video camera. One night, after he and his wife had treated an FBI colleague to dinner at their home, Currie and the agent decided to head down to the basement to take a look at the latest data from the investigation.

That's when they saw a flurry of traffic going into and coming out of Mafiaboy's residence. Currie and the FBI agent immediately thought they had another denial-of-service attack on their hands. That was a possibility the agents had been facing all along. Figuring out how to conduct an investigation while at the same time trying to prevent another round of attacks was a big task.

Currie yanked a few of the data packets from the stream and made a live copy to analyze. If you know what to look for, you can learn a lot from the raw data packets. If it's HTML, or Web traffic, you can tell that. And although it's more difficult, you can also tell if it's e-mail. Ten minutes passed, and Currie's anxiety grew. Then, all of a sudden, they noticed data packets containing messages such as "I'm going to kill ya," "Death God," and the like. Mafiaboy wasn't in the midst of another denial-of-service attack against major e-commerce Web sites; he was playing an online game called Starcraft, a real-time strategy game that pits three races against one another in an intergalactic war.

Then Currie watched him tinker with some of the hacker tools he had used in the original attacks in February. But just when the teenager looked like he was getting back on track with his hacking activities and possibly starting to learn something, Currie noticed, on March 21, that he had launched a limited ICMP attack against himself. Kids. They never seem to learn.

Mafiaboy's ineptitude didn't surprise investigators. School had never been high on Mafiaboy's priority list. Classmates and school administrators describe the computer whiz kid as somebody who had been repeatedly suspended for discipline problems. In fact, prior to his arrest, Mafiaboy had reportedly been suspended twice from Riverdale High School. After his arrest, he would even violate the terms of his bail by getting suspended upon his return to school. Classmates and teachers recounted incidents where the teenager talked back to his English and math teachers and banged his fists on his desk out of frustration. He rarely showed up for class with his books or with completed homework assignments. Mafiaboy had a real attitude problem, one fellow student said in April after the hacker's arrest.

Such an attitude problem didn't fit in with the culture of Riverdale High School, an ethnically mixed school of about 1,200 students. Mafiaboy preferred to dress in baggy pants, baggy jacket, and Nike tennis shoes, and he was often seen wearing a baseball cap in the backward punk style of many teenagers. In contrast to those who said he was a normal kid, other friends said he hung out with the tough kids at school, smoked cigarettes, got a lot of play with the girls, and was generally a troublemaker.

Mafiaboy was a fish out of water in a school with the motto "Reach Higher and Succeed." Riverdale wasn't his first school, of course; he had been thrown out of his first school because of discipline problems. And unlike at his first school, at Riverdale, students were required to wear uniforms. Forest green and black were the school colors. White oxford shirts, white turtlenecks, forest green cardigans, V-neck pullovers, and blazers were the uniforms of choice. No sneakers, running shoes, black jeans, black Palazzo pants, sweatshirts, or boots of any kind were allowed.

These details would not emerge until after agents had taken Mafiaboy into custody. Prior to that, however, the RCMP was learning other critical details about Mafiaboy's home life. The teenager's choice of hacker nickname was no accident, Knesek would later recall. "He didn't pull the name Mafiaboy out of the air."

⋆^⋆ ( ) ⋖ [ ] ⋗ ( ) ⋆^⋆

It was April 15, forty-three days into the wiretap and data interception operation, and a clearer picture of Mafiaboy had emerged. The wiretap proved to be the critical tool in the investigation that enabled investigators to link Mafiaboy to the technical forensic evidence. His guilt and the fact that he had acted alone had also been well established. The RCMP still needed, however, to be absolutely positive about who was sitting in front of the computer.

By this time, Mafiaboy had turned 15, and his older brother recently celebrated his eighteenth birthday. This was good news as far as Gosselin was concerned because if it turned out that the older brother was responsible, he was now an adult and could be charged as an adult. Gosselin and Knesek had pictures of the family, but the ages of the brothers were close enough that it was difficult to tell who was talking on the telephone. Sometimes the agents had to listen closely to what the boys were saying to figure out which brother was talking. They had similar likes and dislikes, and both talked about girls.

They talked about Mafiaboy, too, and that proved to be a key piece of evidence pointing to the younger brother. In addition to capturing the teenage hacker's voice talking about the fact that he had conducted various hacks, the investigators also captured his older brother bragging to friends about his younger brother's hacking exploits. At one point, the brother bragged about how his little brother was all over the news, a clear reference to the February denial-of-service attacks.

John Calce, Mafiaboy's father, also found his son's hacking accomplishments impressive. But the businessman would rather have avoided the type of attention the attacks had brought, according to investigators. The 45-year-old transportation company executive had other problems, and the attention of law enforcement was not what he needed at the moment.

Calce's immediate plans involved hiring a hit man to assault a business associate because of a dispute involving a $1.5 million business transaction. Calce would later receive the equivalent of a slap on the hand for what Gosselin and Knesek feared could have been a plot to commit murder.

For 43 days, Gosselin had resisted the temptation to storm in and confiscate the teenager's computers and had every intention of continuing the wiretap for the entire 60 days he had been authorized to run it. But now the RCMP had evidence that the boy's father was conspiring to really hurt somebody. The plan was all set. Tonight's the night the two men could be heard agreeing on the telephone. Investigators had to move in. "We didn't want anybody to get killed," recalled Gosselin.

Police raided the house at three o'clock in the morning on April 15. However, all they found was a surprised and bewildered father, the stepmother, and Mafiaboy's two brothers. Mafiaboy was nowhere to be seen. They took the father into custody and were informed that Mafiaboy was staying at a friend's house. When RCMP agents arrived at the friend's house, Mafiaboy was standing outside on the curb, fully dressed and relaxed. He looked like he was waiting for a bus or hailing a taxi. From the look on his face, it was exactly what should be happening. It was time to go in, and he knew it.

Although Knesek was not present when the raid was conducted, she recalls the wiretap and a portrait of a highly dysfunctional family. There were padlocks on the doors to each of the brother's bedrooms. Mafiaboy "saw a lot, dealt with a lot, took a lot," recalled Knesek. Neither Mafiaboy nor his father considered what Mafiaboy was doing illegal or harmful, says Knesek. "That was the way the family lived." And although Mafiaboy did begin to earn a little respect as a hacker, it lasted only until other hackers became too afraid to be around him.

^ ( ) < [ ] > ( ) *^*

With Mafiaboy in custody, the RCMP investigation that had kept five investigators busy for 60 to 80 hours a week and cost $100,000 (U.S.) had finally come to a close. The Mounties got their boy. But when investigators picked apart the teenager's computers, they found no technical evidence linking him to the attacks. Mafiaboy's hard drives and any other evidence he may have had lay somewhere at the bottom of the Lake of Two Mountains or one of the many other lakes, rivers, and tributaries that weave in and out of the

Montreal area. Without the wiretap and the original evidence captured by UC Santa Barbara administrators and others, the Mounties would not have had a case.

Mafiaboy pleaded guilty in a Canadian youth court to dozens of charges related to the February attacks. The one charge he refused to plead guilty to, however, was the attack against Outlawnet in Oregon. I didn't do it, said Mafiaboy. The RCMP withdrew the charge and to this day suspect that one of Mafiaboy's brothers was responsible.

But pleading guilty was all Mafiaboy did. He wasn't talking. Gosselin tried repeatedly to interview the teenager to find out why he did what he did, what his motivations were, if there had been anything pushing or forcing him to conduct the attacks. But the boy and his lawyer, Yan Romanowski, repeatedly refused. The one and only time that Gosselin and other investigators had a chance to interview Mafiaboy, Romanowski was present. The Montreal hacker had decided that he was going to take his chances and hope that the court would believe his contention that on February 7, 8, 9, and 10, and again on February 12, he was simply running tests that would have enabled him to design and build a new and improved firewall device.

There were a few holes in Mafiaboy's story, however. First and foremost was the fact that his so-called tests lasted for six days. In addition, the hacking tool he had downloaded and used came with an explicit warning that it was illegal to use the tool against another computer network, and that it was not designed to collect statistics or information that could be used to build a new firewall. His story was completely bogus and another indication of Mafiaboy's lack of sophistication. "He wanted to see if he could take these big companies down," recalled Gosselin. "The evidence proves that without a doubt. There was no way out for him."

Mafiaboy appeared at his sentencing hearing in June 2001 wearing baggy pants and a blue dress shirt, untucked and sloppy. The final blow to Mafiaboy's defense was delivered by the court-appointed social worker tasked with interviewing the teenager and his family. The social worker told the judge that "not only is he not taking full responsibility for what he did, he's still trying to justify that what he did was right." A 16-page report submitted by the court expert concluded that Mafiaboy had lied when he said that he was only trying to test the security of the Web sites he attacked. If that were true, argued the social worker, the attacks wouldn't have lasted as long as they did.

The court-appointed social worker later recommended to the judge that Mafiaboy receive five months in closed custody for his crimes because the teenager posed a moderate risk to hack again. Mafiaboy's mother responded to her son's prosecution by telling the judge that she felt she might have been too strict on the boy when he first showed signs of an obsession with computers, but that his father was not strict enough in supervising and guiding him. Parents have a responsibility to guide their children, she told the court.

Although a defense criminologist testified that Mafiaboy had clearly taken responsibility for his crimes and had accepted his guilt, the prosecutor, Louis Miville-Deschenes, used reports from teachers and school administrators who knew Mafiaboy to paint a picture of a disrespectful, unruly troublemaker who craved attention. That was the Mafiaboy that Swallow, Gosselin, and Knesek had come to know. And who better to know Mafiaboy than the agents who watched and listened to his every move?

On September 12, 2000, a Canadian judge slapped Mafiaboy with an eight-month sentence in a juvenile detention center. The maximum sentence he could have received was two years. The judge also prohibited Mafiaboy from possessing any software not commercially available and banned him from using the Internet to talk with other hackers and hacking into any other Web sites. He also ordered Mafiaboy to tell authorities the name of his Internet service provider.

^ ( ) < [ ] > ( ) *^*

From that day on, the mystique of Mafiaboy ceased to be a force in the online world. In less than three months, he had gone from being a bold, self-proclaimed ruler of cyberspace who dared authorities to catch him to a kid whose scrawny frame and unrefined technical skills were unable to withstand the undertow of the law. Many would argue that the teenager was nothing more than a bumbling amateur, unworthy of the attention he received. But his record speaks for itself.

The last time the FBI had dedicated nationwide resources and manpower to hunt down a single hacker was during Kevin Mitnick's online rampage in the late 1980s and early 1990s. It was a sign of the times, however, that the Mafiaboy investigation involved more than a hundred agents in two countries, compared to the three agents who worked full-time on the Mitnick case.

The similarities between the two hackers are striking, though. Neither was a very good hacker. Mitnick was far better at social engineering than he was at hacking, although he was technically advanced for his time. Mafiaboy, while relatively good at marketing and hyping himself online, routinely struggled to use basic hacking tools and techniques that other hackers had long ago mastered. But neither hacker needed to be exceptionally skilled at actual hacking. Mitnick's success was a direct function of the phone numbers, account numbers, and passwords he was able to fool unsuspecting people into giving him.

And for Mafiaboy, successful hacking was a matter of downloading software tools from the Internet, following directions, and hitting the Enter key on his keyboard. Mafiaboy wasn't about finesse and technological discovery; he was about crime. He was a pretend hacker who gave hackers everywhere a bad name.

Ironically, Neal left the FBI for a job with Exodus, the same company that had initially denied his agents access to the company's facilities to investigate leads in the Mafiaboy case. Neal eventually took Swallow and Knesek with him to form the foundation of Exodus' specialized Cyber Attack Tiger Team. As of the writing of this book, Gosselin and Currie remain on the beat with the RCMP in Montreal.

All of the investigators who took part in the hunt for Mafiaboy have stated for the record that they have absolutely no doubt that the Montreal teenager acted alone. The evidence from the wiretap and data interception operation is overwhelming, they say. However, corporate security experts and other members of the hacker underground say the real architects of the February 2000 distributed denial-of-service attacks have yet to be found. Mafiaboy didn't have the skill to pull off such a highly sophisticated and coordinated series of attacks, they argue. More important, the tactics used during the attack against Yahoo! differ significantly from the tactics used during the attacks against the other sites, say some corporate security experts who were involved in deciphering what happened. And if that's not enough to convince people that he didn't act alone, they say, there's also the high volume of false confessions and the ease with which some hackers were able to dupe others during IRC chat sessions into believing that they were in fact the real Mafiaboy.

Neal, however, remains confident that the Mounties got their boy. "I was right in the middle of this case," said Neal. Allegations that Mafiaboy didn't act alone are "absolutely false." And on the issue of Mafiaboy's guilt: "There's absolutely no doubt."

4

A Tale of Two Script Kiddies: Pr0metheus and Explotion

Be not hasty in thy spirit to be angry: for anger resteth in the bosom of fools.

Ecclesiastes 7:9

They are brothers in hacking, similar in as many ways as they are different.

Pr0metheus is angry with Christianity. Explotion is just angry. Pr0metheus channels his anger into his hacking, acting as the Devil's advocate within the electronic underground. Explotion rages against the system, unsure how, where, or against whom he will direct his anger in the future.

Both Pr0metheus and Explotion started hacking during their early teenage years. And although events have taken them in different directions, a fine thread of youthful angst and resentment binds them to one another.

There are many like Pr0metheus and Explotion. Few are understood. Some, such as Pr0metheus, don't care what labels are applied to them as long as their message gets across. Others, like Explotion, acknowledge the labels applied to hackers by the so-called experts, but resist being labeled themselves. But for both of these hackers, the act of hacking is more a sideshow than a goal. In many ways, their hacking has taken a backseat to other personal motivations and viewpoints. Some might read these stories and conclude that hacking is being used. Others, while they may or may not agree with Pr0metheus' and Explotion's personal views and opinions, may accuse them of hijacking the hacker ethic and tarnishing the reputations of all hackers.

Regardless of what you decide, all hackers, even script kiddies like Pr0metheus and Explotion, have unique stories to tell. These are the stories of the teenage hacker who would become The Devil's Advocate and the teenager who continues to Rage Against the System while trying to figure out what type of hacker he should be.

^ ( ) < [ ] > ( ) *^*

The Devil's Advocate

His name is Pr0metheus, and he's one of Satan's hackers, sent from the molten depths of the lake of fire on a search-and-destroy mission. His targets: Christian Web sites.

It's November 20, 2001. Pr0metheus is in planning mode, conducting reconnaissance for his first wave of victims on the Internet. He opts for speed. Servers running the Microsoft Windows operating system will be the easiest to deface quickly. Then he chooses his exploit; tonight he'll take advantage of a common configuration error made by many system administrators who run Web sites using Microsoft's FrontPage Web server software. He knows that most administrators of these systems set the access privileges incorrectly, allowing "everyone" to alter, delete, upload, or download information on the server at will.

The first thing Pr0metheus needs is a list of victims to attack. For this, he uses the Web site netcraft.com, an online service that allows anybody to search for and locate Web sites containing certain words in their domain names. Pr0metheus searches for all hosts that contain the words *church* and *holycross* in their names. In seconds, the online system spits back a list of more than 2,000 servers.

But it's impossible to deface 2,000 Web servers in one night, even for a minion of the Devil. Pr0metheus has to narrow down his target list. To accomplish this, he's written or modified several automated scripts using the Perl programming language. Perl, an abbreviation for Practical Extraction and Report Language, is designed especially for processing text and is the perfect choice for writing scripts that will deface Web servers. The best script kiddies learn to master Perl early in their careers.

Pr0metheus runs the script to narrow his search on the first list generated by the keyword *church*. It looks something like this:

C:\script.pl contains bible mylist.txt

The search turns up only a few dozen Web site names that also contain the word *bible*. That gives him a little more than 100 servers containing the words *church, bible, holycross,* or a combination of two or more.

The next script checks only for those servers that are running the Windows operating system. Now the list is narrowed down to a manageable number of potential victims. Finally, he runs the last script, which checks for FrontPage systems that have open ports and access controls that enable "everyone" to modify the site's content. Pr0metheus cracks a menacing smile as the final target list is generated. There are plenty to choose from. There's nothing his victims can do to save themselves now. And because all of the systems are running their Web services on an open port, port 80, not even a firewall can stop Pr0metheus.

One of his first targets is the Southgate Baptist Church in Springfield, Ohio. It is a Web site that "emphasizes a relationship with God as the central focus that brings all of life together." It stands for everything that Pr0metheus does not. He is a worshiper of Satan, a lowly pawn of Lucifer crawling on his belly through the Internet in search of new disciples of the father of all lies. Defacing Christian Web sites is not only part of Pr0metheus' war against Christianity; it's a method of recruiting and a twenty-first-century means of spreading the false word.

With a few clicks of his computer's mouse, Pr0metheus sets the script in motion that replaces the Southgate Baptist Church Web site with his own liturgy on the principles of Satanism. Tomorrow morning, when the parishioners of Southgate visit the church's Web site in search of words of inspiration and hope, they will be confronted instead with Pr0metheus' message, which he types in blood-red text set against a black background:

owned by
Hacking For Satan

I. Satan represents indulgence, instead of abstinence!
II. Satan represents vital existence, instead of spiritual pipe dreams!
III. Satan represents undefiled wisdom, instead of hypocritical self-deceit!
IV. Satan represents kindness to those who deserve it, instead of love wasted on ingrates!
V. Satan represents vengeance, instead of turning the other cheek!
VI. Satan represents responsibility to the responsible, instead of concern for psychic vampires!

VII. Satan represents man as just another animal, sometimes better, more often worse than those that walk on all fours, who, because of his "divine spiritual and intellectual development," has become the most vicious animal of all!
VIII. Satan represents all of the so-called sins, as they all lead to physical, mental, or emotional gratification!

When he's done authoring his eight pillars of the church of the damned, Pr0metheus offers unsuspecting Web surfers a link to the frequently asked questions page of the Church of Satan. He's hopeful that this will be enough to entice believers into a life of moral relativism. Pr0metheus even leaves a corporate calling card: his Hacking For Satan e-mail address:

hfs@godisdead.com

The defacement of the Southgate Baptist Church Web site takes only minutes to complete. Then Pr0metheus moves on to his other victims. He hits the Web sites belonging to the Mountain Ridge Regional Church in Pennsylvania, the Alliance Bible Church in New Jersey, the Crosstown Community Church in Florida, and the Holy Cross School in New Zealand. All of the defacements are the same; except for the one that targets the Alliance Bible Church. For that defacement, Pr0metheus replaces his eight-pillars of Satanism with a picture of an upside down cross and a quote by author Richard Lederer:

There once was a time when all people believed in God and the church ruled.

This time was called the Dark Ages.

It's that simple. A new defacement group called Hacking For Satan is born, and Pr0metheus is its leader. The initial phase of operations, phase one as Pr0metheus calls it, continues for the next three days. More than a dozen other religion-oriented Web sites will be defaced with the same message.

But Pr0metheus gets creative from time to time. His hack of the Web site biblequestions.com includes a graphic depiction of a Satan-like creature dressed in a shirt and tie authoring a document titled "One Soul." Just as Satan understands the value of one soul and is willing to pay any price to capture it, Pr0emtheus understands the value of a single defacement and is willing to take any chance to pull it off. Pr0metheus is Satan's personal hacker. Both are ravenous in their appetites.

Pr0metheus' defacement of www.biblequestions.com, Nov. 22, 2001

On November 22, Pr0metheus defaces the Web site called Topical Bible Studies. In the center of the page he places a symbol of a goat's head in the middle of a star, known to Satanists as the Sigil of Baphomet. There are five Hebrew letters at each point of the Baphomet: Lamed (L), Vau (V), YOD (Y), Tau (Th), and Nun (N). These letters spell the Hebrew word *Leviathan* (LVYThN), which is associated with the Ouroboros, or serpent eating its own tail. The serpent represents the Big Bang of the cosmos, the simultaneous creation and destruction of the universe. The serpent devours his tail as his body expands. The star comes from the Pythagorean school of mathematics and to Satanists is associated with another form of the serpent, Leviathan. The goat is Pan, the symbol of nature.

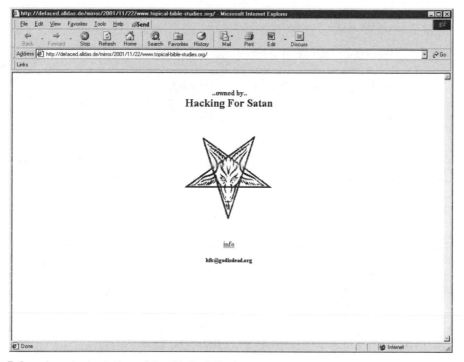

Pr0metheus's depiction of the Sigil of Baphomet

To Pr0metheus, these three elements represent the Trinity of the Church of Satan, the three clans or schools of thought united under one symbol. The so-called saints of this false religion are bizarre: Dr. Jack Kevorkian, Madonna, comedian George Carlin, and ultra-feminist Camille Paglia, among many others.

Pr0metheus wants nothing more than to spread the word of Satanism. He views his Web site defacing activities as a sort of "hactivism"—activism through hacking. It's not about raising peoples' awareness of Internet security issues, and it's not about the love of technology or the thrill of the hack that devotees of the true hacker ethos seek. It never was and never will be. He does not feel evil. Pr0metheus simply hates organized religion, especially those denominations that fall under the umbrella of Christianity. Hacking is just one way to strike back.

On November 30, 2001, the Archdiocese of Baltimore issues the following bulletin:

Church Web Site Hacker Alert

We have been made aware that a group called "Hacking for Satan" is targeting church websites worldwide, replacing the legitimate site with satanic material and links. They are currently active on the east coast and have disrupted 30+ sites in the last week. All parishes with websites should be diligent in reviewing their sites to ensure that they will continue to be secure.

There will be many more victims. Nobody is safe. The era of Hacking For Satan has only just begun.

^ ( ) < [ ] > ( ) *^*

Pr0metheus started hacking when he was 14, about the same time he discovered Satanism. But his parents were more concerned about the phone bills generated by his online activities than they were about his choice of religion. The only time his father ever really hit the roof was when the phone bill arrived. Soon he asked the telephone company to restrict the duration of outgoing calls from the house to three minutes. Until Pr0metheus figured out how to hack his way around the limitation, he was known in the hacker underground as the "Three Minute Caller."

Pr0metheus's hacking began in the usual way: with a combination of game cracking and hacking into his high school's network to cheat on tests and assignments. It was a Novell network that used mostly Microsoft Windows workstations throughout the school. One day, Pr0metheus and a friend got their hands on a list of default passwords through a creative social engineering campaign. With the default passwords at their disposal, they were free to roam the network, looking through the personal accounts of teachers and fellow students alike. Naturally, they created hidden accounts for their own use and granted themselves root access rights. Pr0metheus learned what it felt like to be God on the network. It was the one time the thought of God didn't repulse him. He located answer keys to tests and copied other students' work.

The school administrator was lame, as were the teachers, when it came to computer security. The computer instructors, especially, barely understood what they had been entrusted to teach the students. Most of the kids, especially those like Pr0metheus, were self-taught and became bored easily with the rudimentary lessons in C and Pascal programming. Hacking was a challenge, a way to show off and wield power. It was fun, too. School was old, tiring, and monotonous.

Pr0emtheus' ownership of the high school network lasted only until his hacker cohort got busted and ratted him out to the school principal. The young hackers were in a bad situation, facing all sorts of potential punishments. To Pr0metheus' surprise, the principal offered him a deal he couldn't refuse. In return for not being expelled, Pr0metheus agreed to provide lessons in hacking and security to the school's network administrator and to the various school computer instructors. For a moment, the roles were reversed. The looks on the teachers' faces were priceless. To Pr0metheus, the situation was more than a little funny. It was absurd.

For a while after that, Pr0metheus drifted away from computers. He had other interests, such as listening to music, playing the guitar, and reading. It was then that his reading habits changed course. His philosophy on life was undergoing a radical transformation, and nobody, not even some of his closest associates, realized it was happening.

^ ( ) < [ ] > ( ) *^*

Prior to his fifteenth birthday and what customarily would have been the year of his confirmation in the Catholic Church, Pr0metheus' parents told him that he was free to choose his own religion. Although his parents both were raised in families where organized religion was the norm, they thought it would be best for Pr0metheus to choose his own belief system. The self-described high school outcast, known for his black wardrobe, unwillingness to conform to the rules, and regular occupation of the computer lab, chose Satanism.

Pr0metheus formed his first opinions about Christianity during his junior high school classes in religion and history. In those lessons, he saw people being enslaved, wars being waged in the name of religion; he saw feminists and homosexuals being persecuted ruthlessly for what were not, in Pr0metheus' mind, sins. But the real transformation happened when Pr0metheus read Anton S. LaVey's 1969 book *The Satanic Bible*.

"It opened my eyes," he recalled years later, referring to what many might call LaVey's mindless, moral bankruptcy. "It had a message that was easy to agree with. Us humans, we are just animals like the rest. We have natural instincts that we need to follow. Religion has been used throughout history to control and suppress people, telling them this or that is sinful, making them feel guilty. I want to rise above [that]."

Pr0metheus' teenage mind devoured LaVey's deceptions. The Satanic master's espousal of a life of self-indulgence that focuses on carnal desires appealed to Pr0metheus, whose unbridled youth was in search of an alternative to what he considered Christian propaganda. "Religion has been at the root of a lot of terrible things in this world," he said. Pr0emtheus was clay in LaVey's hands. Satanism was the kiln that solidified his future belief system.

^ ( ) < [ ] > ( ) *^*

Pr0metheus' organized hacking career started not with Hacking For Satan, but with another well-known group called PoizonB0x. Responsible for more than 900 Web site defacements, including many of the mass hacks that targeted Chinese government sites during the May 2001 cyberwar between U.S. and pro-Chinese hackers, PoisonB0x offered Pr0metheus the perfect learning environment.

One of his first Web site hacks targeted the Defense Enterprise Computer Center, a St. Louis–based component of the Pentagon's network provider, the Defense Information Systems Agency (DISA). On May 24, 2001, Pr0metheus hacked the DISA site using the same exploit he would use months later against various church Web sites. He replaced the DISA page with the message "PoizonB0x Wuz Here."

On May 25, Pr0metheus hit NASA's Langley Research Center, once again replacing its Web page with the message "PoizonB0x Wuz Here." The next day, he broke into the Web site belonging to the Neutron and Nuclear Science department of the Department of Energy's Los Alamos National Laboratory, the home of the U.S. Defense Department's nuclear weapons research. He left behind a message that was a variation on a theme: "PoizonB0x Owns Jooo."

Pr0metheus' first online defacement rampage made headlines. Trade journals and newspapers ran stories about "cybervandals" having turned their attention to government and military Web sites. More than 400 Web attacks were recorded each day between May 15 and June 1. PoizonB0x claimed responsibility for many of them.

In June, Pr0metheus attacked the Web pages of the Washington State Court system. Later that summer, on August 11, he returned to the Los Alamos National Laboratory. This time, however, he taunted the system administrators about the lack of security at their Web site and left behind a picture of a bald Anton LaVey. Pr0metheus' Satanism was beginning to bleed over into his hacking.

Pr0metheus' work as a member of PoizonB0x, Los Alamos National Laboratory Web site defacement

Things were changing for Pr0metheus. He was getting bored with PoizonB0x, especially the method and the message. Life as a PoizonB0x member was getting tedious. He wanted to deface Web sites for a cause, not because it was easy. Pr0metheus felt like an inner-city punk spraying graffiti on the side of a railroad car. He was yearning for something more.

Pr0metheus also missed the personal interaction and comradeship he remembered from his days in junior high school when he was cracking games. Most of the well-known hacking groups, such as PoizonB0x, are loose-knit organizations with members all over the world. The majority of the hackers in PoizonB0x are from the Baltic countries, and the interaction level among members is minimal. Pr0metheus recalled, "Contact with other members wasn't very personal at all. You basically got to cover your own ass and couldn't really trust anybody out there." Within six months of joining PoizonB0x, Pr0metheus decided that it was time to move on, time to give his hacking a purpose.

Pr0metheus claims to have contacted a "few friends" to start Hacking For Satan, but the secretive nature of a group that remains actively involved in mass Web site defacements makes confirmation impossible. He may operate alone; nobody knows for sure. What is clear, however, is that Pr0metheus gets a rush, "a sense of accomplishment," out of defacing Web sites with Satan's message. He enjoys "outsmarting" the administrator of the sites he attacks.

But now it's about more than outsmarting the administrators that run the Web sites. Now his hacking has a purpose, regardless of how misguided that purpose may appear to others. Pr0metheus hacks for Satan and to spread the word of Satanism. "It is more satisfying now," he says, "because I hope to get more people interested in Satanism. If it wasn't for that, I would have quit defacing when I quit PoizonB0x."

^ ( ) < [ ] > ( ) *^*

It's January 6, 2002, forty-seven days since the birth of Hacking For Satan. Pr0metheus has just finished defacing the Web sites of Jesusfire.com and the St. Mary's Schools in Annapolis, Maryland.

"Pray if you like to kneel, if you like to lay. The war against Christianity continues," he wrote in the defacement of Jesusfire.com.

"Ironic," he said, in his message to the administrator of the St. Mary's Schools Web site, referencing the warning sent out by the Archdiocese of Baltimore.

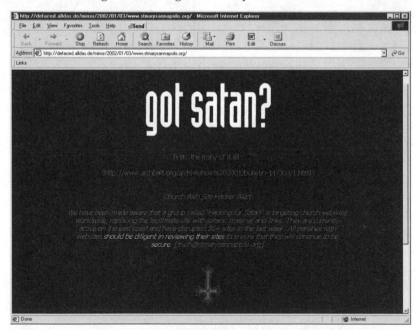

Pr0metheus' defacement of the St. Mary's Schools Web site, Jan. 2, 2002

With the addition of these latest casualties, Pr0metheus' victim list stands at 56 Web sites and counting. Most are Christian sites, but there's also one called the deepsouthjewishvoice.com, which requires a user name and password to access, a Salvation Army Web site in Korea, and a Sanyo Web site in Istanbul, Turkey.

Pr0metheus has no plans to stop hacking. In fact, he's busy working on "big plans" for the future of Hacking For Satan, including an expansion of the group and campaigns other than Web site defacing. He has no fear of getting caught.

"I have an ethic when I deface Web sites that I follow 99 percent of the time," he says. "I never destroy any data on the servers, save for log files when necessary. I back up original files, and I do not copy existing files to my own computer. As long as I don't steal credit card information, I don't find it very probable that someone will do all the necessary work, with a small chance of finding me, and then press charges against me. But of course, there is always the risk when involved in criminal activity. I am aware of it, and it would seriously affect my future career should I get caught."

His future career? That's right. In addition to working part-time in a record store, Pr0metheus is in school studying—you guessed it—computer security.

Pr0metheus looks at his past and current activity as a so-called script kiddie as "not very flattering," but nonetheless a phase that he, like most hackers, must go through to learn about computer security. Although he fits the definition, Pr0metheus feels he has evolved since he "first earned the title."

"We [in Hacking For Satan] use exploits and scripts that we modify or write ourselves," he says. "And, of course, I don't use an exploit that is meant for Unix against an NT host. But you have to realize that every hacker label has been stigmatized to a certain extent. So I don't really care what people call us, as long as our message gets out there."

And that's what it's all about for Pr0metheus. The hack isn't the message. "We deface Web pages to promote Satanism, not to improve security, not to show how advanced our skills are, not to earn some abstract form of respect or anything like that," he says. "We will continue to attack and deface Christian and other religious Web sites that we find appropriate. Every such site should feel threatened."

^ () < [] > () *^*

Rage Against the System

He goes by the online handle Explotion. A few years ago, he misspelled the word *Explosion* while registering for a hacker Web site and thought it looked cool that way. He thinks of it as a concatenation of the words *explosion* and *motion.* More important, nobody uses the name Explotion. Uniqueness has always been an important factor when choosing a hacker handle.

Explotion is 19 years old and pissed off. He's not angry with anybody in particular; he's angry in a general sense. He doesn't like idiots. And most people are idiots. The fewer people Explotion is forced to deal with, the better. After all, nobody enjoys "knocking heads with 10,000 idiots a day." From the lamers who are constantly asking him how to hack, to the bum on the street who's always asking him for money and the "[f***ing] assholes" who blow cigarette smoke outside of his apartment window, most people are the same to Explotion. They're idiots.

But Explotion is not like most people. He's a hacker, and a pretty good one. He's smart, well educated, and can find his way in and out of a computer wearing a blindfold. Sure, he's angry. But a lot of hackers, especially teenage hackers, are angry. Explotion understands the anger. The script kiddies, hackers, whatever you want to call them "have their reasons," he says, for expressing so much violence, anger, and, sometimes, even hatred. It could simply be the way they talk, or the only way they know how to express their frustration, he says. Or it could be for the shock value and nothing more. Who's to say for sure? Different people and different generations speak different languages.

Curse words are considered obscenities only because our parents told us not to say them, according to Explotion. He gets a kick out of watching MTV bleep out the curse words in music videos that reference "good things," at the same time playing the uncut versions of songs like "Macarena," which makes references to giving your body pleasure and to a young girl who cheats on her boyfriend by offering to take his friends home with her.

"I've always wanted to play a song about the destruction of the world and the raping of little children," says Explotion. And you know what? "As long as I don't cuss and I make the song catchy, people will play it for their kids."

^ ( ) < [ ] > ( ) *^*

Born and raised in California, Explotion was a product of a bitter divorce that occurred when he was only five. He lived with his mother, a brother, and a sister. They were constantly on the move, shuffling from apartment to apartment. It was as if they were always running from something, he recalled.

Explotion couldn't remember a time when computers were not part of his life. His father, a 30-year veteran of PacBell who always had a computer, taught Explotion how the telephone network worked and helped spark his interest in technology and the Internet. His older brother, on the other hand, took out his frustrations physically on Explotion. His sister preferred to manipulate their mother, Explotion said.

It was around the time that he entered junior high school when Explotion began to see "the wrong in his parents and the wrong in everybody else." His mother and father argued in court about money. His father usually lost and would be forced to send half of his paycheck to wherever Explotion and his mother were living at the moment. Explotion thought of his mother as a "grandma," who would "rather lie to you because she thinks it will make you happy in the short run." His father, though, was always straightforward. Explotion recalled the time that his father admitted to him that he wished he'd "never had kids." He always told Explotion the truth, even if Explotion got to see him only on weekends. "I respect his honesty though," Explotion said.

Explotion needed an outlet, an escape from the chaos. So he turned to the things that interested him most: he played guitar and cello. He also delved deeper into computers and, of course, hacking.

Hardly a day went by during junior high school that Explotion didn't hack into somebody's Web site. Sometimes he did it for revenge, sometimes for fun or to actually help a classmate with a programming glitch. Eventually, he was hacking into his teacher's machine, copying answers to tests, and distributing them to his friends.

For two years, Explotion studied Cisco networking. The course was offered by his high school. Explotion and a friend, also a hacker, finished the four-semester course in less than two. They spent the next year using the school network to figure out things that the Cisco course didn't teach students.

One of Explotion's favorite hacking pastimes was a hacking competition sponsored by his computer instructor. Students would each get five minutes to secure their mini-networks so that nobody could break in. The first person to shut down every computer on another student's network won. Explotion won often. But even when he didn't, he learned something because the winner was required to share the tricks that enabled him or her to succeed.

Although the best students in the class won most of the competitions, everybody benefited. Explotion's Cisco instructor understood this. Rather than move at the pace of the slowest student and alienate the most gifted hackers, Explotion's teacher recognized the opportunity to tap into the talent of the best hackers in the class. She used kids like Explotion to provide the less skilled students with real-world examples of security and hacking in action. Regardless of whether you won or lost, you always learned important lessons about the technology, hacking techniques, and the best way to protect a system. The live hacking sessions complemented perfectly the traditional classroom instruction.

But Explotion's Cisco teacher wasn't only cool; she was also the school's network administrator. Explotion and his friends actively searched for security holes on the network and eagerly reported them to her. They even started the school's System Administrator Club, where uncovering security holes, patching them, and helping to administer the network became their job. Hacking became one of Explotion's extracurricular activities, and he excelled at it.

One day, Explotion uncovered a major security gap that allowed anybody with the right set of skills to break into students' e-mail accounts. He had been poking around on the network, administering his own account, when he realized that by manipulating a few hidden variables, it was possible for him to bypass the account password. During another incident, which can best be described as a semi-denial-of-service attack involving an anonymous e-mail message that replicated itself throughout the network, Explotion managed to trace the culprit to a system inside the school. The hacker turned out to be one of Explotion's friends, so he didn't report his findings. He did, however, block his friend's e-mail account. You have to be a system administrator to understand such cold-hearted logic. Friend or no friend, there was no way Explotion was going to go through *that* every other day.

^ ( ) < [ ] > ( ) *^*

In high school, Explotion was one of those kids who didn't get teased a lot, but who also wasn't cool enough to be popular. He was sort of nobody, hidden, walking in the background. People knew who he was, though. He wasn't invisible. He played soccer and water polo. But he often found himself on the outside looking in, disagreeing with the status quo, and not getting along with the other students.

During his junior year, Explotion left public school and enrolled in an experimental technology high school. Funded by Hewlett-Packard and about 40 other technology firms, New Technology High School in Napa Valley, California, focused on teaching students about—what else?—technology: computers.

Only the best of the best got to attend. Explotion had taken a test a few years earlier, the results of which demonstrated his technical prowess and ability to "figure things out." There were only about 100 juniors and 100 seniors in the entire school. But the place was rigged for hacking. The school purchased 145 HP Vectra VL Series 4 desktop computers, and 140 HP color monitors, plus HP LaserJet and DeskJet printers, HP ScanJet scanners, HP AdvanceStack hubs and switches, and HP DeskDirect 10/100VG PCI LAN adapters. The Internet was as common as a yellow number 2 pencil.

Explotion was happy to be there. At New Technology High, he could be in front of a computer all day, every day. Plus, there were fewer idiots to deal with. Unlike his old school, he could get up and go to the bathroom whenever he needed, or slip out to the local donut shop down the street. Suddenly, his clothes didn't determine his social status, nor did his athletic abilities or how big of an ass he could make of himself. Here, his social status was based on his knowledge: his knowledge of computers. He was treated like a hacker.

There were still instances, of course, when somebody would knock heads with Explotion. This was to be expected. It was, after all, high school, and Explotion's attitude toward other people wasn't exactly what you would call pleasant. But revenge took on a decidedly technical tone at New Technology High. Those who decided to pick fights with Explotion usually arrived the next day to find thousands of e-mail messages flooding their in-boxes, an obscene e-mail addressed to the school administrator logged in their Sent folder, or, Explotion's personal favorite, a picture on their personal school Web portfolio home page showing their face on a fat guy making love to a horse.

^ () < [] > () *^*

A few months into his junior year, Explotion was living with his mother in a trailer park close to his new school. Surprisingly, he met another kid who had the same interests: music and computers. The kid lived two trailers away.

Explotion and his new friend would sneak out of the house every night and stay out until the bus arrived in the morning to take them to school. Explotion's mother worked nights as a respiratory therapist, so it wasn't a big deal for him. The only person he needed to avoid was his older brother.

But for Explotion's friend, running around town all night was an escape. His mother had died when he was 13, and his father liked to practice his left jab on his face. Eventually, the boy's mother's friends adopted him after he'd made several attempts to run away. He liked to call his adopted parents Grandpa and Grandma to their faces, but to Explotion, he referred to them as Grandpa and Bitch. Explotion recalls how his friend fantasized about killing "the bitch." But nothing ever happened, fortunately for everyone. Explotion's friend talked a bold game but rarely acted upon his frustrations.

A young girl lived in the trailer between Explotion's and his friend's. She was okay, but nothing special. If you were away from home for the first time and met her, you wouldn't write to your mother about her. But if you were into computers like Explotion was and you met her father, you might want to strike up a friendship. The girl's father worked for some company as a computer security specialist and owned more than a few high-speed computers. Eventually, Explotion talked his way into his next-door neighbor's house and was granted access to the computers to do whatever he wanted with them.

Soon, the girl's father was bringing home computer parts for Explotion. Within two months, he and his friend had gathered enough parts to build their own system. They kept the system at Explotion's friend's house because he wasn't allowed to spend the night away from home.

Explotion left home the day he turned 18 and moved into an apartment with a new girlfriend, close to his school. It wasn't unusual for his family. His brother had lived with a girlfriend when he was only 16 and was married and divorced by 19.

Explotion was close to graduation by this point. Living on his own was better than living with his mother and brother. He had managed to acquire a computer thanks to a part-time job he had at a local visitor's bureau. The city had given away a bunch of old computers, and Explotion had grabbed one that he soon turned into a Pentium 133-MHz system with 96 megabytes of memory. It worked well enough.

As he approached graduation, Explotion received a notice in the mail that said he wouldn't receive his diploma until he returned an overdue physics book he had checked out of the school library. It wasn't that Explotion had forgotten to return the book; he wanted to keep it. It was one of the better books on physics, and he hadn't had the chance to finish it during class. Not returning the book would lead to a $70 fee, the letter stated. Not to worry: it was a simple hack, as easy as grabbing answer keys to tests. In fact, it was the same hack that Explotion had conducted many times before through another teacher's system. Once the record was changed and the log files cleaned up, nobody could prove that the book hadn't been returned. Things disappear in the virtual world all of the time.

$$\star\,\hat{}\,\star\,()<[\,]>()\,\star\,\hat{}\,\star$$

It's October 2001. Explotion is out of school and enjoying his new role as a chief technology officer at a Web development company near his home in California. It's a small company of about five people. Explotion recently hired one of his hacker friends to work as his assistant. He still doesn't like people, though. The commotion and noise of the four other people working in his company's three-room office space gets to be too much for him. So he moves himself into another room, where he can work alone, and an over-sized corkboard helps keep his desk free of clutter.

Explotion is finally making some real money these days, which makes his hacking more enjoyable and effective, especially since he can afford a better computer. He recently bought a new system with a state-of-the-art processor and video card and a dual-boot hard drive running both Windows 98 and Red Hat Linux.

Explotion does most of his hacking and experimenting from work. And in the spirit of a hacker who understands that hacking isn't only about computers, Explotion hacks his work environment from time to time. He makes sure that there are certain aspects of his company's operation that only he understands. But the art of the hack is secondary to the outcome. Hacking is simply the tool.

He directs all of the programming and manages the company's Web server, network, and network security. Any of the Web page development that the president and vice president don't want to do falls to Explotion, because he's faster. He likes his job but would prefer eventually to find a gig that would allow him to experiment more. He's a hacker in his heart, and experimentation is his passion.

One day, Explotion was sitting in his office with his assistant, kicking back and surfing through the Web site of the heavy-metal band System Of A Down. The loud L.A.-based hard-rock band was taking a poll on its Web site, where fans could vote for their favorite song from the band's new album. Explotion's assistant had a definite opinion about which song should be the top song in the poll, so he voted for the song and began to refresh the Web page at a rate of about two dozen times a minute. Each time he hit the Refresh button on his browser, the Web site registered another vote for his favorite song.

To Explotion, this was no way to go about fixing an online vote. A hacker's skin crawls at the thought of manual data entry. And in this case, hitting the Refresh button was manual. That's what scripts are for.

"If you really want your song to get more votes, we should write a program to do that," Explotion explained.

Explotion wrote a short, yet complicated, looping script. He used PHP, which stands for PHP: Hypertext Preprocessor, a cross-platform HTML-based embedded scripting language. The script added a vote each time it ran. And since it was a loop, when the last line of the program ran, it simply commanded the program to run again.

The script worked like a charm. But Explotion's mind started to wander. The possibilities seemed endless. He was flying through the deep recesses of cyberspace, where there were dozens of new worlds to explore. What if you could create your own form and use a text box to vote for any song you wanted to rather than limiting yourself to only the songs that the Web site offered? Explotion was not only hijacking an online poll, he was injecting new criteria into the competition. The poll was being conducted by his rules now.

With the use of his own form, Explotion began voting for three songs that the band had cut from the album. He had downloaded them from the Internet before the final album was released anyway. Word of what he had done got around through online chats that Explotion's assistant had with other hackers and friends on IRC. That's when the prodding started. Everybody wanted Explotion to add obscene titles of songs that didn't even exist. But Explotion only wanted to make the point that the band shouldn't have cut those three songs. That was enough for him, and he soon began searching for a way to contact the Web site administrator to let him know of the vulnerability.

But hacking the site was easier than finding contact information for the administrator. Explotion couldn't find anything, not even an e-mail address or telephone number. So he did what any hacker with a conscience and a sense of responsibility would do. He posted a message on the site informing the administrator of the problem and how to fix it. That seemed like a perfectly logical thing to do, and helpful. He liked System Of A Down and certainly wasn't out to sabotage their site. The administrator for the Web site eventually did make the fixes, but not before other hackers had copied Explotion's hack and created their own forms and began registering votes for songs with titles such as "Talk About Bad Coding" and "Weak Security."

Since then, Explotion has been careful about what he does online. Anything more serious than the System Of A Down hack he won't talk about. Talking sends even the best of them to jail, he says. Also, if you get caught hacking any system today, regardless of how minor the incident, "you're considered a murderer. Nobody will hire you. And the government won't let you near a computer. So the chances of working on one all day are small."

Although he's not big on labels and hasn't decided yet where he fits in the overall scheme of hackers, Explotion does recognize three different groups in the hacker scene today: crackers are the hackers with malicious intent, hackers are people who want to learn, and lamers are posers who want to be crackers without learning. "Hackers are good for computer security," he says. "Crackers are bad for computer security, and lamers are harmless and give up."

There's no telling where Explotion will end up or what road he'll take. In the meantime, he tries to learn everything he can. He dabbles in game programming as a way to satisfy his urge to explore, watches movies, drinks a lot of coffee, exercises every now and then, practices hacking on legal hacking Web sites, and drives around town with his friends. From the outside, there's nothing peculiar about Explotion. But on the inside, there's a war being fought. On the inside, he's still raging against the system.

The teenage hacker "Mafiaboy," who was accused of disrupting the operations of some of the biggest companies on the Internet, including Yahoo.com and CNN, enters a Montreal courthouse on January 18, 2001, where he pleads guilty to most of the 67 charges against him.

RCMP agent Marc Gosselin receives a commendation for his work on the Mafiaboy case from FBI legal Attaché M. Stuart Sturm.

The RCMP team who broke the Mafiaboy case (from left to right): Sylvain Roberge, Marc Gosselin, Michael Haring, Mike McCrory, Patrick Boismenu, Robert Currie, Jean-Pierre ROY (in charge of unit at that time), and Sylvain Aubry.

Anna Moore, 15 years old, at the University of Oklahoma, where she simultaneously earns high school and college credit through an independent study program.

Anna, seen here wearing her DefCon jacket, once tried her hand at phone phreaking using pay phones like this one at a local mall.

Joe Magee at the annual DefCon hacker conference in Las Vegas. Within six months of owning a computer with a modem, Joe went from knowing almost nothing about computers and hacking to knowing everything there was to know.

Joe Magee never really left the hacking scene; he simply changed the definition of what a hacker is. He now works as a computer security consultant. And what he sees taking place in the hacking world doesn't impress him. "If you're not hacking for scientific and educational reasons, then you're doing it for money," Joe says. "Why else would you do it? Credit card data is cash. And as far as Web site defacements are concerned, how does that help anybody? You're not going to get a job if you hack into a site."

Despite having no electricity or running water at home, "Genocide," seen here at the age of 12, would become a skilled hacker by surreptitiously using his mother's access to a local university network.

"Genocide," seen here at the age of 17.

5

World of Hell

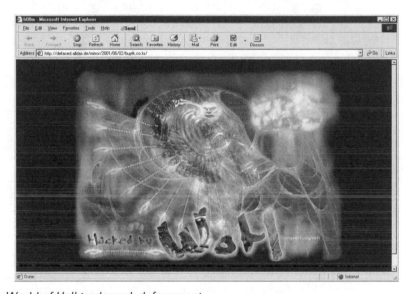

The World of Hell trademark defacement

Hackers are not the big, bad evil people like most of the general public thinks they are. News programs and magazines are starting to get better about how they talk about hackers, but still about 80% of the people (especially those that can't turn on a computer) are the ones that are terrified of hackers. And now that Sept. 11 has happened the government is getting more money and feeding on the fears the general public has about hackers and using it to arrest some of the most intelligent people in the world, or anyone that could stand up to them when they're wrong. They just say "they're hackers," and right then most of the population hates you cause of their bad perspective of hackers. Then the government says that what you're doing is a major threat to the economy and well being of the country (which is a load of bullshit) and you get thrown in jail and no1 cares :/ So, to all the people that think all hackers are evil = wise up. And to all hackers = be careful, big brother is watching.

Cowhead2000, founder of the World of Hell group

After the September 11 terrorist attacks on the World Trade Center and the Pentagon, Congress granted federal law enforcement authorities sweeping new powers to tap telephones and track suspected terrorists on the Internet. There was even talk of treating hacker incursions into government computers as terrorist acts. It was the hacker underground's worst nightmare come true. Hackers were now in the same category as Osama bin Laden's ruthless murderers.

It was 5:30 P.M. on Tuesday, November 27, 2001, exactly 11 weeks after the September 11 attacks, when the feds raided the Memphis, Tennessee, home of the 15-year-old founder and leader of the infamous hacker gang World of Hell. It was a raid unlike any he might have anticipated, had he anticipated one: no sound of doors being kicked in, handcuffs clanking around wrists, or shouts of Miranda rights being read; just a surreal, hyperventilation-induced light-headedness that told the teenager's brain that the situation was serious.

His underground name was Cowhead2000, and he never imagined he would be in this position; he was too good, took too many precautions. For the first time in his hacking career, however, he was genuinely scared.

Cowhead2000 had founded World of Hell in March. Its members, known to be located all over the world, quickly earned the collective reputation as the undisputed kings of mass Web site defacements. The gang was famous for defacing dozens, sometimes hundreds, of Web sites at one time. The total number of sites "owned" by the WoH crew, as the denizens of the hacker gangland knew them, was in the thousands. The victim list included Web pages from private companies, U.S. and foreign government agencies, and military organizations.

It wasn't until the federal agents arrived at her doorstep that Cowhead2000's mother had any idea that her son was involved in hacking. She had complained dozens of times before about his 12-hour stints sitting in front of the computer. But she thought he was like every other kid, obsessed with the wonder of the Internet, chat rooms, games, and of course, the easy availability of porn. In her mind, hacking never entered the equation. And computers seemed like a welcome distraction for her son, who recently was kicked out of high school for possession of acid and pot.

The agents explained that they only wanted to "talk." Nobody was going to be arrested, they said. Not this time, anyway. They did, however, need to discuss some very serious issues with Cowhead2000, a known associate of some of the underground's most wanted: RaFa, dawgyg, r00t, FonE_TonE, and others. According to one of the agents, after September 11, they were required to follow up on all threats made against the government or government computer networks. There was a real chance that terrorists would try to attack important U.S. government computers, possibly as a diversion to help distract attention from a more deadly attack. And it was necessary for the agents to determine if a recent threat attributed to a member of World of Hell was just a teenager pulling a prank or a serious plan in the making.

Once the initial shock wore off and the teenager's mother calmed down, the agents asked to see his setup. Cowhead2000 showed them to his room. The feds stood there for a moment, astonished by the sheer volume of high-tech gadgetry that lay before them. Cowhead2000's bedroom wasn't a bedroom at all; it was a combination high-tech museum and hacker laboratory. It was a hacker's asylum.

Cowhead2000 took a few minutes to explain to the agents what they were looking at. He did all of his work at the L-shaped desk in the corner, he said. On one side was a 15-inch NEC liquid-crystal display monitor, and on the other side was a Hewlett-Packard DeskJet printer. Beneath the desk sat an HP Pavilion computer with an 850-megahertz Pentium III processor. It was a dual boot system, he told the agents, capable of running both Mandrake Linux 8.1 and Windows 2000 Professional operating systems. Next to the NEC display was another monitor that was connected to a motherboard he had purchased for $20 in an auction on eBay. The motherboard was sitting on a small table next to the desk and was powered by an AMD K6-2 333-megahertz processor and ran version 4.3 of the FreeBSD operating system.

It was a lot of computing power to comprehend, but Cowhead2000 wasn't done yet. Not even close.

On the floor, between the small table and the desk, was a 1.4-gigahertz AMD Thunderbird system running Windows 98. It led to another 17-inch monitor and was used mainly to play games. To the right of the Thunderbird system was an old Compaq machine that Cowhead2000 rarely ever turned on, and next to that was a Silicon Graphics Indigo 2 running Irix version 6.3. Next to the SGI box was a Sun Microsystems Sun Sparc workstation running SunOS Unix version 4.1, and next to that was a Sun Sparc 2 that also ran SunOS 4.1. The Irix box and the Sun workstations were running headless, but next to them sat an older Packard-Bell monitor, just in case he needed it.

In the opposite corner of the room was what Cowhead2000 referred to as his junk pile. There was an old PacBell 9000, which, Cowhead2000 explained, had an ancient version of Windows loaded on it. Next to that was an HP Vectra that until recently had been running Corel Linux. One day the box just stopped working, so he took it apart and threw it in the corner.

There were other bits and pieces of computers lying around, but the ones he had just pointed out were the main systems, he said. That was all of it: nine computers, eight operating systems, four monitors, and a printer. There was barely enough room for a bed. That didn't matter, though. Cowhead2000 didn't sleep much.

The agents appeared genuinely impressed. They commented how it would be nice to have the same type of computing power at their federal office. Then one of the agents spotted Cowhead2000's guitar case and asked him what kind of guitar he owned. Cowhead2000 took the guitar out of the case and showed the agent a white Lonestar Fender Stratocaster with a white turtle shell–design pick guard. The agent asked him all sorts of questions about the guitar and said he was planning to buy his nephew one just like it for Christmas.

It wasn't long before the small talk ended. Cowhead2000 was wondering when they were going to get to the point. Finally, one of the agents asked him about a member of World of Hell named RaFa, who had recently made public comments about targeting military Web sites and taking down the U.S. government. In addition, RaFa had recently shown some people on the Internet an e-mail he had received from Russian hackers looking for help in an alleged credit card fraud scam.

Cowhead2000 told the agents that he knew of no such plans, and that RaFa, whoever RaFa really was, was speaking for himself.

RaFa was, in fact, a 17-year-old member of World of Hell. He was one of the outspoken members of the gang and was known for dabbling in politically charged Web site defacements and enjoying the media attention he got from it all. RaFa had also led a group called the Dispatchers in a campaign to attack and disable Internet connections within Afghanistan and Palestine. In the wake of the Sept. 11 attacks, the Dispatchers had defaced the Web site of the Iranian government's Ministry of the Interior and hundreds of other organizations throughout the Middle East. They also crashed a few Palestinian-controlled Internet service providers. To the Dispatchers, the attacks were more about self-defense against religious fanatics than they were about patriotism.

Also linked to the Dispatchers was another known member of World of Hell: a 17-year-old hacker who went by the underground nickname dawgyg. Like RaFa, dawgyg had been bragging to reporters about a plan to track the Internet use of Osama bin Laden's al Qaeda terrorist network. He also claimed that the group controlled more than 1,000 computers around the world that could be used to launch attacks against the terrorist group's communications network.

Another hacker, who also had links to Cowhead2000 from an earlier group called Ph33r-the-B33r, then stated publicly that the Dispatchers had the power to knock an entire country off of the Internet for a week, and that Osama bin Laden should not take their threats lightly. All they had to do was try hard enough, the hacker said.

Although the threats were coming from a bunch of teenagers, the FBI's cybercrime arm, the National Infrastructure Protection Center (NIPC) in Washington, D.C., took them seriously and at their word, especially the possibility that the group controlled thousands of computers surreptitiously. That was a problem officials had feared for years. In fact, senior members of President Bill Clinton's National Security Council had held several secret meetings during and after the Year 2000 date change to discuss the threat from malicious code and so-called zombie computers. The fear was that terrorists or foreign intelligence agents might have penetrated foreign firms that were involved in contract work to fix U.S. computers for the Y2K glitch and been in the position to implant logic time bombs in tens of thousands of computers around the country. It was a matter of national security. And since then, all claims of "owning" thousands of systems for possible use in future widespread attacks had to be investigated, whether those claims were made by terrorists or teenagers.

Cowhead2000 stuck to his story. He denied any knowledge of RaFa's plans and refuted having any links to the group called the Dispatchers. If RaFa was planning anything big, like taking down the government or establishing a link in the U.S. for Russian criminals, it was his own plan and something he never discussed with the rest of the group, Cowhead2000 told the agents.

That was the truth, of course, but in the underground the truth is a chameleon that changes colors as often as hackers change IRC nicknames. World of Hell guarded its secrets as closely as any intelligence agency did. Its members had gained entry not by bragging, but by proving their willingness and ability to pull off hacking crimes and then fade into the shadows of the underground. The group operated similar to a terrorist cell: they communicated through private, invitation-only chat rooms; nobody had met any of the other members face to face; and they each exercised tight control over what the others knew about them. Few knew each other's real names, ages, or locations. And that was the way they preferred to keep it.

It was 7:45 P.M. when the agents got to the last of their questions, including a few about some of Cowhead2000's recent escapades with a server belonging to a U.S. embassy system. They had little to show for the two-hour discussion. But the agents did manage to accomplish something. By the end of the conversation, they had firmly established the identity of the hacker named Cowhead2000, determined that he was not, in fact, acting in support of a terrorist organization, and had effectively scared the defacer out of the teenager.

"Look, we know what you've been doing on our networks," one of the agents said as they began to leave. "If you keep hacking our systems, you're going to get into real trouble."

With those words, the agent ended the Web site defacing career of the hacker known as Cowhead2000. The meeting had, in the teenager's words, "scared the shit out of me." The defacing scene no longer seemed worth the risk. It was lame and full of kids who didn't want to learn. The WoH gang had become a bunch of "media whores," in his opinion. It was time to get out while he was still on top.

In his mind, Cowhead2000 knew the scene was dying. But in his heart, he knew that the World of Hell would live on. There were others who were ready, willing, and able to step into the leadership position. And there were dozens more who just wanted a chance to be part of the group.

^ () < [] > () *^*

If you want to pin down a day when it all started, you have to go back to March 7, 2001. It was Cowhead2000's fifteenth birthday, and he decided, out of boredom really, to hack a Web page and deface it with a happy birthday message to himself. The victim was Sewon Teletech, a manufacturer of power amplifiers in South Korea.

The defacement of Sewon Teletech was easy, almost too easy. Cowhead2000 had been hacking for about three years, starting out slowly with simple Windows hacks, exploiting well-known unicode bugs in Web server software, and experimenting with the Sub7 trojan on his friends' computers. From there, he moved on to legal hacking sites, such as progenic.com and cyberarmy.com. He also read every hacking text he could get his hands on. But in the few short years he had been hacking, Cowhead2000 had managed to break his way through every skill level offered by the ethical hacking sites. Those sites simply weren't hard enough any more; they offered little challenge and excitement.

Eventually, he banded together with a bunch of friends he had met on various IRC channels, and they formed a hacking group called Ph33r-the-B33r. The group defaced a few dozen Web sites before some of the members split off to form another group called Hackweiser. But Cowhead2000 was interested in finding a real challenge. He wanted to experience the thrill of defacing hundreds of sites, and not only those belonging to no-name companies; he wanted to get to the Web sites that he liked to call "the big boys." Cowhead2000 wanted to own the Internet. And there were at least seven other hackers who were willing to help him make a down payment and take some of the credit along the way.

One of the first Web sites to fall victim to World of Hell was, not surprisingly, a porn site. The defacement was an introduction of sorts, a press release sent to the rest of the underground. It warned the online community that the seven hackers known as the World of Hell were not just another kiddie group. To the contrary, World of Hell was a new group of professionals who was to be "ph33red"—feared—by all. They wanted 2001 to go down in history as the year that nobody was safe from the World of Hell.

A few days later, on March 15, the group hit half a dozen sites simultaneously. The defacement was the same one they had used against the porn site, only this time World of Hell was recruiting new members. "We're looking for coders, no script kiddies please," the message said.

The e-mail messages poured in, more than 20 a day for the next week. The vast majority of requests came from wannabes who "couldn't hack their way out of paper bag," Cowhead2000 recalled a few months later after quitting the scene for good. Only those hackers with an established track record of defacing and who were clearly qualified were allowed to join immediately. Everybody else was given a quota of sites that they had to hack before they would be allowed in. "You had to actively contribute to the group. If you did only five defacements and then didn't do anything for about three weeks, you got kicked out." The wannabes always quit anyway.

On March 25, the group pulled off what would become its trademark: a mass defacement of Web sites belonging to companies and organizations in Mexico and Russia. When the dust settled that night, 100 sites in Mexico and two dozen in Russia lay in ruins.

By the end of March, World of Hell had grown to 13 members and was on the warpath. The defacements piled up. Web sites belonging to companies as wide ranging as Sony Semiconductor Foundry Services, the State University of New York at Stony Brook, the Egyptian government's Ministry of Communications, a Time Warner broadband Internet services page in Wisconsin, and the Hong Kong Education Department fell victim to the group.

World of Hell continued to rack up the defacements through April and May. The group targeted dozens of sites, hitting a few high-profile organizations, such as the State of Virginia Bankruptcy Court, a page belonging to 3Com, and a bunch of servers running everything from Solaris to Linux to SCO Unix and Irix.

Surprisingly, the only thing the group did at these sites was change the content of the index page and erase the logs that would enable somebody to trace the attack. Few, if any, of the attacks involved destruction of information. It was a point of pride for the group, and part of their philosophy. World of Hell was out to show the world that every Web server was hackable. And by doing that, the group was helping to plug holes in servers run by administrators who didn't know the first thing about security. The reasoning was simple: it's better that companies get defaced by World of Hell than infiltrated by stealthy criminals who would undoubtedly take advantage of the same vulnerabilities but carry out much more serious crimes. Cowhead2000 recalls how the group justified its hacks:

> WoH was about having fun. And we figured that if we defaced a box, yeah, there would be downtime and maybe a little money lost, but what if we or someone else who hacked the box didn't deface it? What if they erased their tracks and backdoored it and kept coming in and using the box for illegal things and taking personal and sensitive information from it and no one would ever know? At least when you deface, they know someone has been there, and they fix it so someone else more malicious can't come in and screw things up.

By June, however, the group's tone was changing. There was a new hacker on the scene named RaFa. He was louder than most of the others in the gang. He used large graphics, issued pointed warnings about the group's invulnerability, and tackled what he considered to be important political and social issues. He became one of the most active defacers in the group, and he considered the name World of Hell representative of the real world in which he lived. It was his mission to use hacking as a tool to strike back at a world gone mad. And more than anything, he was hacking as part of a defensive battle against the violation of the principles of the online world by the morally bankrupt physical world. RaFa hated the system, and he was hell bent on changing it.

^ ( ) < [ ] > ( ) *^*

Born a hacker in 1984, RaFa grew up somewhere in Middle America, suburbia—and that's about all he'll say. He's unwilling to provide any more details on his early childhood because that's how crackers, especially the ones that are wanted by the FBI, get caught, he says. "If you're going to get involved in illegal shit, you need not talk about it."

RaFa was always a smart kid who got good grades in high school. He mostly kept to himself, at least until his senior year, when he busied himself with "pimping a lot of women," to use his words. But hacking was always a part of who RaFa was. For him, hacking was never about a label or being cool. It was a way of thinking, an instinct, a sixth sense. You either have it, or you don't. And you don't get it from the media.

Graphic arts was another of RaFa's passions. He chose the hacker handle RaFa after the artist Raphael, whom he considered to be the greatest painter of the Renaissance period. But what made Raphael an attractive role model for the young hacker was the fact that the Renaissance artist was a genius on a par with Michelangelo, and yet he lived like and among common men. Raphael is best known for his Madonnas and the large compositions that decorate the Vatican in Rome. RaFa would make good use of these works in some of his Web page defacements.

One of RaFa's first defacements for World of Hell was a mass defacement of more than a dozen Windows servers belonging to relatively obscure companies around the world. Although not one of the more difficult or complex hacks he would conduct, RaFa's message set the stage for what became the beginning of his campaign to make a difference in a world that appeared beyond all hope. RaFa wrote:

> *Can I live in a world where young children and women die on their way to school and work?? Every day when I take the bus I see a man with a suitcase and I think what will happen on my way to school? Will I die?? The nations are the reflection of their government rulers and their unemployment, ignorance, sub development. It's hard to survive in a world with limitations! How many people will die? When will all this shit be over? People from Brazil and Venezuela are dying for hunger. For this reason we live in the World Of Hell....*

RaFa wasn't the average World of Hell member. He was different, outspoken. His was the voice of a generation disgusted by the world that its parents, grandparents, teachers, community leaders, and politicians had created. People didn't feel safe in their own homes any more. Nobody could be trusted. A person's word didn't mean anything. Truth had lost its battle to the cancerous growth of corruption that now infected every corner of society. People talked about giving the system a chance to work, but to RaFa, "the system sucked."

World of Hell, on the other hand, was tight. The members had never met each other face to face, but they trusted each other and considered one another close friends. "We have each other's backs," RaFa once said. "In this world, you need to have allies."

It was June, and RaFa had met an ally in dawgyg, one of the first members of the World of Hell gang. Dawgyg was 17 and a varsity wrestler and soccer player who was as adept at failing classes as he was at hacking. He left another group called ph33r hax0r crew to join World of Hell after Cowhead2000 asked him if he was interested in defacing a few sites for them. He was a prolific defacer who wrote most of his own exploits. With a click of the mouse, his programs dropped him in on dozens of Windows and Unix systems with root access privileges—access to the main directory. And he, too, was beginning to use the media attention to give a voice to his stand on social issues.

On June 9, dawgyg hit two Solaris-based Web servers belonging to the Commonwealth of Virginia, including the Commonwealth's home page—its face to the world. He used the defacements to attack what he called the "favoritism" practiced by college Reserve Officer Training Corps (ROTC) programs and the fact that "patriotism is dead in the youth of America."

Dawgyg hit more state government sites later in the week, prompting newspapers and trade journals to run headlines about hacker gangs that were waging war on state government computers. One of the sites, however, was the home page of the Texas State Lottery Commission's Bingo division. That hack, which occurred on June 13, raised serious concerns about the potential for tampering with the lottery and led to a thorough scrub of the systems for signs of a more serious security breach.

On the same day that he hit the Texas State Lottery, dawgyg also upped the ante further, defacing the Web site belonging to the National Petroleum Technology Office of the Department of Energy. This was a federal site belonging to an agency with a long history of serious security breaches involving alleged misuse of computer access.

The hacks of government Web sites continued for the next five days. Dawgyg took credit for more than a dozen such defacements, taunting the administrators about the lack of security on their systems. He discovered more vulnerable systems than he could attack during any one day. And as far as dawgyg was concerned, the administrators who were responsible for securing the computers deserved it. After all, he was using exploits for which patches had been made available more than a month earlier.

But as in the past, there was no indication that dawgyg did anything other than deface the Web pages of the servers he gained control of. There were no reports of stolen credit cards or machines being ransacked. Although damage was done, the only files ever really deleted were the log files that showed who logged in, at what time, and from where. Most of the defacement messages dawgyg left behind said something about the need to improve security and offered a pointer to the original Web page.

For dawgyg, World of Hell had a simple mission: "to show that even the big boys can be hacked." Everything can be hacked. Nothing is unhackable, he said. It simply depends on how much time you spend hacking it. Patience and persistence were the two most misunderstood virtues of the hacker underground. Sometimes you simply needed to take your time.

RaFa, for example, knew a thing or two about taking his time. Some of his hacks took him weeks to accomplish. Others took minutes. Sometimes he used code that he wrote himself; other times he used programs that he downloaded from the Internet or got from another member of the group. He tried everything he could find. But he was never all that impressed by what he was able to do, including setting a record in July of 679 simultaneous defacements. "It's a script. Nothing special," he recalled. "My best hacks are the ones nobody knows about."

^ ( ) < [ ] > ( ) *^*

World of Hell's expertise wasn't limited to simple exploit scripts. Unlike most teenage hacker groups, this one made a point of recruiting hackers with a wide range of skills and knowledge. Everybody was a programmer to a certain extent, but some members were clearly more advanced than others. One such member was a 17-year-old named FonE_TonE.

FonE_TonE was a former member of a group called r00t-access and had been hacking since he was 15. He started with the basics, such as network enumeration, domain name system interrogation, network reconnaissance techniques, and various other trace-route tactics. But then he became interested in routers and started to conduct simple scans using the Internet Control Message Protocol (ICMP) to ping, or test, a connection and locate network gateways. During these early days of discovery, he often forgot to use proxies to mask his true identity; the administrators of the networks he was surveying could have easily traced the scans back to FonE_TonE, but they never did.

FonE_TonE's hacking knowledge increased at a meteoric rate during the first two years of his hacking career. But unlike the stereotypical teenage hacking whiz kid, FonE_TonE didn't need to sacrifice his personal life to achieve his hacking goals. He did well in school and had a lot of friends, and he acknowledged that he always avoided hanging out with the "skanks," the uncool crowd. At times he was a rebellious, difficult student, who was good with the jokes. His fellow classmates voted him "Most Humorous" during his senior year, in 2002.

But behind the laughter lived a kid who at one time or another during the course of his teenage years belonged to an underground gang of cybervandals. The only person who ever knew about his life in the underground was his best friend. To his parents, he was a dabbler in network security, Web page design, programming, and cryptography. To them, World of Hell—whatever that might be—was another parent's problem.

FonE_TonE first rose to prominence in the hacking and defacing community when he published two informative hacking texts that taught less experienced hackers some of the finer tricks of the trade. One of the texts explored the importance of deleting log files on target systems and how the files are created and where to find them in different operating systems. It was a text written, as he said, for more advanced "hax0rs."

"I wouldn't recommend this to any of you people who have just figured out what the 'ls' command does in a Unix system, or just found out what TCP/IP means," FonE_TonE wrote. "This is basically how hackers get caught. If you do not know how to use the [log] cleaners then you shouldn't be in the system."

Of course, one well-received hacking text deserved another. This time, he wrote about a topic of intense interest for him: Cisco router security. The paper described common vulnerabilities that had been announced to the public a year earlier and instructed hackers on how to find companies that had not yet patched their systems to close the hole. However, in pure FonE_TonE fashion, he also discussed how easy it is for administrators to fix certain holes in their systems. After all, improving security was what hacking was all about. FonE_TonE didn't hack for the sake of hacking; he did it because he was passionate about learning about security and spreading the wealth of knowledge he gained during his hacking expeditions.

"I have so many freaking routers that are 'owned' by me that it's not even funny," FonE_TonE wrote. "Actually, it is."

His second text also tackled the more complicated issues surrounding tactics for defeating encryption and how to determine whether a router is running something called Trivial File Transfer Protocol (TFTP). If a router is running TFTP, FonE_TonE wrote, the hacker simply needs to download the configuration files for the router. It was a complicated subject, but FonE_TonE never left his audience in the dark without offering references for additional information. In this case, he directed readers to a Web site that should be familiar to everybody who has read Chapter 1 of this book: "More info at this site I found --> http://www.Genocide2600.com."

"I hope you have learned a little something on how idiotic administrators are," wrote FonE_TonE. "So please be careful. I hate having to find retarded shit, especially the default passwords. I just now counted 34 routers with default passwords on them. I have already changed each router's daily banner telling the administrator to be more careful."

^ () < [] > () *^*

A short time after he published his hacker texts, FonE_TonE approached the leader of r00t-access and asked if he could get in on the action of defacing Web sites. The front man for r00t-access, known by the nickname grimR, read FonE_TonE's texts on deleting log files and router security and welcomed the brash kid into the group with open arms. The indoctrination was simple: find Web sites and deface them.

After a handful of defacements targeted at relatively obscure Web sites such as a university in Australia, FonE_TonE decided it was time to try to get in with a bigger and better crew. At this point, everybody wanted in with World of Hell. However, FonE_TonE had several advantages, not the least of which was a small, but growing portfolio of Web hacks.

Cowhead2000 assigned him a quota of at least five hacks to do in a week. Not only did FonE_TonE surpass the minimum requirement with ease, but he produced what remain to this day some of the most elaborate and graphically stimulating defacements of the group.

For example, on July 25, he defaced the Web sites of office supply giant Staples Online, the Shaolin Kung-Fu Academy in British Columbia, and the home page of online catalog vendor Prostar Interactive MediaWorks with a doctored photo of a knight wearing chain mail and flanked by lightning bolts set against a deep blue sky.

FonE_TonE's defacement used against Staples Online, July 25, 2001

As other members of World of Hell did, FonE_TonE made a point of leaving his e-mail address along with a note to the system administrator of the Web sites he attacked offering his assistance with security. Sometimes he even told the administrator what exploit he used to compromise the system. In FonE_TonE's mind, gaining root access to corporate computers and defacing Web sites was about the legitimate pursuit of knowledge and security.

"We like to find new things, see what we can do and what we can't do," he said later. "And what we can't do, we try very hard to do. Not all defacements are political, but it's still good to know that we do it for a reason. I hack because I love to learn new things about network security."

^ ( ) < [ ] > ( ) *^*

All hackers love to learn, but not all hackers love to learn in school. That was especially true of World of Hell's kr0nograffik, known by his abbreviated hacker nickname kr0n. He'd been hacking since he was 11. And now, at the age of 15, kr0n wasn't interested in anything that school had to offer, which, he said, was very little.

His high school computer teachers "don't know shit and they never will," he said recently. "They grew up in a generation built on drugs and rock and roll. Why should they know anything? Their lives are almost over, and most do fine where they are."

But Kr0n is happy where he is, too. He was invited to take part in what is known as an "actel," or academically intelligent school, but he declined. And although his teachers were always telling his parents that he was an academic leader, his report cards always included remarks such as "capable of achieving more" and "not living up to potential."

One day, his gym teacher gave him and the other students in the class a pep talk, telling them that they were capable of achieving anything they wanted as long as they worked hard enough. "I meet a lot of average people," the teacher said. "And who wants to be average when you can do better?"

It was a rhetorical question. But Kr0n raised his hand. Heads turned. Laughter filled the gym. Everybody thought it was a big joke. Everybody accept Kr0n. He was dead serious.

"I'm an average student, and that's all I want to be," he recalled months later.

But when it came to computers, Kr0n was anything but an average student. Like all good hackers, he read hacker texts and books about security incessantly. In fact, that's all he did for about a year before he tried to hack his first box.

His first hack involved a Windows system that had one of the NetBIOS shares open. NetBIOS, or the network basic input-output system, is built-in software that contains code for running PCs on a network. He poked around the file system and found a bunch of personal family pictures that belonged to whoever used the computer. When he finished looking through the contents of the system, he reformatted the hard drive. Whatever was on that computer, including the photographs, was lost forever. "It made me feel good," he recalled.

Kr0n also has said that he knows what it's like to feel unsure, paranoid. That, after all, is the essence of life in the underground. The constant fear of getting caught and the uncertainty of not knowing exactly with whom you are dealing online conspire to make even the most experienced hackers paranoid, even the members of World of Hell. And while this world is a secret society based on virtual trust and putting your hacking where your mouth is, the true hacker underground that Kr0n has come to know is a world of shadows, where nothing is what it seems.

Hackers are no longer the geeks that the media paints them to be. They're the bag boy at your local grocery store, the girl standing behind the counter at the mall, the star running back on the high school football team, and the greasy-faced teenager pumping gas into your car. They live double lives full of secrecy. It's a way of life that kr0n fully accepts. "Take me, for example," he said recently. "I've played soccer for nine years. I play football and most sports. I have many friends. I grew up a normal kid with normal friends. But I only tell them what I want them to know or think about my hacking."

And Kr0n has been good at keeping his hacking a secret. Few people know the full extent of his involvement in the underground, especially his membership in World of Hell. He acknowledges that his mother "freaks" whenever the Internet service provider calls with complaints of hacking incidents originating from their house. Of particular concern are the calls from "big brother," which in the past have led to a few heated arguments with his father. But that will never stop him. Hacking is too much fun. In fact, having fun and learning about computer security is what World of Hell is all about as far as Kr0n is concerned. And yet it's more than that. In many ways, being a part of World of Hell is beside the point. For Kr0n, hacking is a way of life, an addiction.

"Hacking isn't something you can just stop doing," he says. "It's a state of mind you get into and you can't get out of it," he recalled when describing his own experience. "You are sitting there ready to take down a major corporation and thinking to yourself, 'I'm about ready to gain access to the site.' You are so into this when suddenly the thought crosses your mind that you can get into trouble. But you don't care. You're about there, and then you have it. I have been tried and found not guilty of several computer and credit card fraud laws. This is not fun. And even though the life of a computer hacker isn't quite cut out as it is in the movies, I still can't stop."

Kr0n may be only 15, but his hacking techniques are advanced. He specializes in *nix boxes, better known as Unix and Linux. Some of the systems he breaks into are so insecure that it takes him only seconds to gain root access. Others prove more difficult and can take several weeks to crack. In those situations, he takes his time finding out everything he can about the Web site, its administrator, and its users. He conducts a thorough reconnaissance, looking for holes, determining where the administrator "lives" on the network and what he or she does. It can be hard, tedious, mind-numbing work. But it's all part of the life and part of the job description that all members of the World of Hell accept when they sign on.

Since linking up with World of Hell last year, Kr0n has hacked many more sites than he's defaced. In fact, he usually doesn't deface a Web site unless he has a particular message that he wants to send to the world. Sometimes he talks about religion or the government in his defacements. On other occasions, he simply sends his "shoutz" to the rest of the World of Hell crew and informs the Web site administrator that the server has been "own3d."

^ ( ) < [ ] > ( ) *^*

December 2001 was a pivotal time for the World of Hell crew. The vast majority of the original members had long ago departed the scene out of fear of being ensnared in a federal dragnet or because they had grown frustrated with what they saw as an underground poisoned by unskilled, mindless kiddies. But all groups need a leader to survive and to keep moving forward. And with the departure of Cowhead2000 from the World of Hell, that responsibility fell to RaFa. And he delighted in the role.

Now, more than ever, RaFa never slept. One workday of hacking consisted of many suns and moons. He gravitated toward his computer throughout the day, trying desperately to live a "normal" life and attend to daily chores. If there was something that he simply had to do and it wasn't related to the computer, he scheduled it between compiling code, installing software, or waiting for a brute-force cracker to complete its task. As hard as he tried to avoid it, the hack was consuming him.

In the waning days of December, it fell to RaFa, and to a lesser extent dawgyg, to keep World of Hell on the map. The defacements continued, including RaFa's very own personal holiday greeting. But internally, the group was dying. The malaise that had infected World of Hell's collective consciousness was taking its toll. Hackers wanted to be a part of the underground's most notorious defacement group without having to actually do anything. World of Hell was slowly becoming like all of the other groups. The promise of media attention was becoming more important than the message.

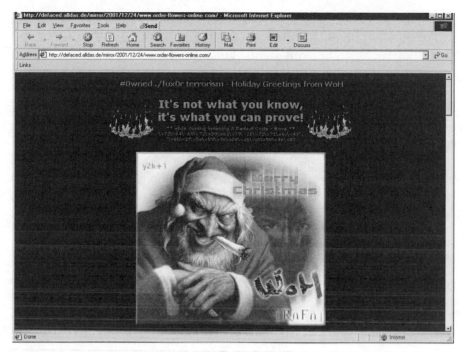

RaFa's World of Hell holiday greeting: defacement of
www.order-flowers-online.com, December 24, 2001

The end came on January 15. That was the day that RaFa signed off for good, leaving a stinging message for the underground and signaling what may be the beginning of the end for the serious Web page hacker community. In his hack of seven sites, RaFa bid farewell.

> *Every man, woman and child has their day in the sun, its how you use that sunlight that matters.*
>
> *I spent a good amount of my life spreading my messages and spiritual philosophies; I tried to pump some intelligence into the underground community. I broke away from the mold of Owning a box for the joy. I have a message. Weather [sic] you like me or not you must admit that I wasn't the usual defacer.*
>
> *I wanted to get the underground involved in political activism. The media and our governments are seriously neglecting the truth, and I wanted to do something about it.*
>
> *I have witnessed the defacing scene become more and more ignorant over the years. Do you want to be forgotten as a kid that just rooted servers, or do you want someone to remember you for changing their life? It's your choice now, god help us.*

-deface scene- *** Notice -- Client exiting: [RaFa] (WoH@leader) [Quit: out]

To the less sophisticated observer of the hacker scene, it was a cryptic message. But to others, it was simple: RaFa was giving his notice. He was quitting.

"The goal in the community should be common," he wrote a few weeks after confirming his departure from World of Hell and from hacking Web pages in general. "That is why I am leaving the defacing scene. They all seem to have lost sight of the real goals. I have to admit that what kept me in it so long was the fame and the friends," he said. "Good times. The journey was long and I learned a lot," he said. "It is my time to go."

The future of World of Hell is in doubt. A new leader has yet to emerge from within the unknown number of other members who remain committed and continue to hack. But one thing is certain, according to RaFa: "The world of hell remains upon us."

6

Cyberchic:
Starla Pureheart

It was a strange, unfamiliar scene to veterans of the hacker underground. DefCon, the dark, irreverent hacker convention held every year in Las Vegas, had gone semi-legit. There were fewer nose rings this year, fewer tattoos, fewer people trying to fill hotel toilets with cement, less black leather, shorter hair, and a lot more people wearing khaki pants and penny loafers. And to top it off, time had been set aside in the agenda for an ethical hacking game called CyberEthical Surfivor (spelled with an *f* to denote the game's Internet roots). It was a contest that mixed ethics with hacker versions of the popular television game shows *Survivor* and *Weakest Link*. That's right; ethics had met the infamous hacker subculture.

There was, however, an overwhelming sense of irony to the event. The objective was not to be ethical, but simply to "win," according to the game's host, long-time DefCon attendee and author Winn Schwartau. "You are required to humiliate people," he said. "This is DefCon." Then there was the presence of Chris Goggans (a.k.a. Erik Bloodaxe, the leader of the infamous gang of cyberpunks known as the Legion of Doom). "I represent the unethical portion of the judges," said Goggans, referring to his participation as one of the game's three judges. Jennifer Granick, the chief judge and a lawyer who in real life specializes in defending hackers, handed out cups of beer as consolation prizes to the losers.

The game had come down to two hackers. Both had survived multiple rounds of hypothetical questions that presented them with various ethical dilemmas. Sixteen others had already been voted out, including a petite, studious-looking blonde teenage girl whom everybody referred to as "Oklahoma."

The 15-year-old had been voted out of the competition several rounds earlier for "being too smart for her own good," as one contestant had put it.

The host asked the final question of the game.

"For children under the age of 16 who do some illegal hacking and cause serious damage, what should we do with them legally, what should we do with their parents, and what is the ultimate recourse that you would design for this country?"

The first hacker gave his answer. "Depending on the nature of the crime, [especially if it] could result in someone's death, the 16-year-old should be tried as an adult...get a year of juvie [juvenile detention], and take cyberethical classes."

The crowd moaned with disappointment at what was clearly a mediocre response.

The second hacker, known as The Doctor, sat on the stage dressed in a black tank top. Pushing a long mane of hair out of his eyes and over his shoulder, The Doctor took the microphone and prepared to give his answer. "I think that the perpetrator should be charged by the severity of the effects of what he had done," he said. "I also think that he should be forced to go to Sysadmin boot camp and should spend time cleaning up the messes of other such attacks. I also think that the youth's parents should be educated."

The crowd again moaned in disappointment. The two hackers had survived long enough to compete in the final ethical face-off, but they clearly were not the ethical knights in shining armor that the crowd and the judges expected. Nor were they articulate.

Granick then asked Oklahoma to return to the stage and to act as a neutral friend of the court. Who better to provide expert advice to the judges on this topic than a hacker who herself had not yet turned 16?

"The parents should be notified," Oklahoma told the crowd. "There is nothing more humiliating to someone under 16, and I know from personal experience, than having your parents find out you've committed a computer crime," she said. "If the parents don't take action, the child should be tried according to the severity of the crime. If it's a crime that causes financial damage equal to grand larceny, then the child should be tried as an adult. Otherwise, they should be tried as a juvenile. End of file."

The crowd of hackers cheered the young girl as she handed the microphone back to Granick and returned to her seat in the audience.

"The judges thank Oklahoma for her opinion and look forward to her ruling someday when she's a judge," said Granick. "But for now, The Doctor wins."

The crowd booed and hissed. The court's decision clearly did not sit well with the audience. Granick's selection was arbitrary. Recognizing the audience's disappointment, Schwartau interrupted and reminded the audience, "Ultimately, you choose the winner."

The audience cheered and announced the winner with a chant. "Oklahoma! Oklahoma! Oklahoma!"

To some, a game where hackers voted each other out of the competition based on their answers to questions about ethics seemed oddly out of place at a convention that was first advertised in 1993 as an opportunity not only to exchange information and meet the celebrities of the hacking underground, but also to enjoy "sheer, unchecked PARTYING." But times had changed.

By DefCon's ninth year in July 2001, conference organizers had been quoted in the media as saying that it was possible to be a hacker and be ethical. There was nothing unusual about DefCon attendees taking part in a game about ethical hacking, they said. And now they had proof. A young, innocent female hacker named Anna Moore had just won CyberEthical Surfivor.

It was a sign of the times. The hacker no longer fit in the male, antisocial misfit pigeonhole. That was an image of a bygone era. To the contrary, hackers were now aware of the consequences of their actions. As a result, they were more apt to be responsible netizens, more apt to err on the side of caution, and unlike many of their predecessors, they were keenly aware of how much the world around them had changed. And if that wasn't enough to shock the veterans of the underground, these hackers were also increasingly female.

^ ( ) < [ ] > ( ) *^*

Born in 1986 to a computer-oriented family in Norman, Oklahoma, Anna Marie Moore was watching her mother work on the computer and learning how to manipulate files and directories in DOS by the age of three. By four, she was reading at a third-grade level and was granted the unusual privilege of using the local library computer system to play educational games without adult supervision.

The year was 1990, and the biggest hacker crackdown to date in the U.S. was underway. The Secret Service, which often becomes involved in crimes involving interstate commerce, had launched Operation Sundevil, and federal agents were raiding the homes of teenage hackers from New York to Texas, including the notorious groups of cyberpunks known as the Masters of Deception and the Legion of Doom. The teens who belonged to these gangs—mysterious dwellers of the electronic underground—were responsible for the Martin Luther King, Jr., Day crash of AT&T's long-distance telephone system. For them, it was a year of living dangerously on the electronic frontier.

But for Anna Moore, these were halcyon days of bright-eyed discovery and learning. She dabbled in educational games, such as Math Blaster and the Mavis Beacon typing program. And when she finally grew tired of those, she moved on to more action-packed computer games like Commander Keen, Duke Nukem, Paganitzu, and Supaplex. She also took the time to learn the ins and outs of Word Perfect and QDOS file manager. All before she had turned eight years old.

It was then that Anna inherited her parents' old computer, fully equipped with a 486 DX-266 MHz processor. Her parents were always upgrading; moving first from an ancient 8086 processor–based system to a 286, then from a 386 to a 486, and finally to a Pentium II. It was when the Pentium II system entered the house that Anna no longer had to share computing time with her mother and father, whose jobs required many hours of sitting in front of the computer. Now she could dedicate real time to her digital explorations, working mostly with early versions of Windows and QBASIC, an older interpreter for the BASIC operating language that Microsoft provided with DOS and Windows.

The computer became an essential part of Anna's daily routine, even before she had her own machine to work with. Her parents had decided early on that home schooling was best for Anna. She was exceptionally bright, and the rigidity of the public school system, her parents believed, would only hold her back. Home schooling, on the other hand, would allow Anna to progress at her own pace. She could study uninterrupted. In fact, she would prove herself to be advanced enough to skip kindergarten and third and eighth grades. Computer programs and online Web sites would play an important role throughout Anna's education.

Unlike most kids her age, Anna's social life did not become intertwined with her academic life. Sure, she made friends in the traditional types of activities in which young girls often participate—Brownies, piano lessons, Christian fellowship, and even tae kwon do lessons—but few shared Anna's passion for computers and technical prowess. "I longed for more intellectual pursuits among peers with a similar level of computer savvy," she recalled years later.

The Internet arrived at Anna's house in 1995. She was nine. Her mother had experience exploring the BBS scene and using telnet software to connect to her office in downtown Oklahoma City, where she worked for a patent attorney.

But that was in the 1980s—a time that Anna refers to jokingly as the era "before the earth's crust cooled." The modern Internet, on the other hand, was a brave new world. And although Anna's parents controlled and monitored her online activity closely by restricting Internet access to their computer, the petite, innocent, self-described computer dork was about to discover her first chat room. It would be there that she would experience her first brush with hacking.

^ () < [] > () *^*

It was the summer of 1996, and during one of her extended online expeditions Anna came across the Unofficial World of Nintendo chat room. Until then, she had experienced only message boards. But this was different. Chat was real-time, and it was exhilarating. She couldn't get enough of it. There never seemed to be enough time on the computer between her parents' work schedules.

By the next spring, however, that problem had been solved. Anna's father had recently built a computer for his own use. Now the Moores were a three-computer family. And it got better. Anna's personal computer, which was located in the privacy of her own bedroom, had a state-of-the-art 33.6K-baud modem. The world was at her fingertips, and best of all, she was outfitted for hacking. "I had the support of my parents, almost unlimited time around computers, and the freedom to explore," she recalled.

That spring, the IRC Nuke Wars started. Hackers, mostly teenagers, began battling for control of the chat rooms. It was the equivalent of a digital turf war. Individuals and gangs of marauding teenage hackers began fighting in the streets of the underground, like a hacker version of *West Side Story*. They kicked each other out of the chat rooms and crashed one another's systems with programs called bots—short for robots—more commonly known as Ping o' Doom or Finger o' Death. The programs were crude denial-of-service attack tools that could be downloaded from the Internet for free and used at will. The kids loved them. And like any curious 11-year-old, Anna downloaded a "nuker," as it was called, and the battle was joined.

The IRC Nuke Wars had a new force with which to contend: a hacker named Star Road. It was a name chosen from the Mario series of Nintendo games that Anna enjoyed so much. She had actually changed her name from Princess Zelda at the request of her parents, who were concerned about the likelihood that a young girl with an Internet handle like Princess Zelda would become the target of every low-life traipsing up and down the information superhighway. So Star Road was her name, and Nuke Wars was her game.

Star Road began nuking other IRC "lamers," as she called them, at will. There was no real point to these battles, except to have as much fun as possible watching your targets disappear momentarily from the chat room. There also was an element of power to the IRC Nuke Wars. If your nuker program could run Ping o' Doom on port 139, a port used mainly by home users for file and print sharing on their Windows systems, you immediately became an object of reverence, "a chat room deity" worthy of being feared and placated, recalled Anna.

But it's hard to be a nuker warrior princess when your parents make it their business to keep a close eye on your Internet activities. Concealment, as futile as that was in the Moore house, became part of Anna's challenge. Although there was no real harm being done, it was never a cool thing when her mother walked into her bedroom as she was running a program that filled the screen with big letters that read, "Nuke 'Em To Hell." Naturally, Anna's parents' interest in her online life became part of Anna's perceived "problem," an obstacle to her explorations. Even worse was the fact that Anna lived in a household where both parents knew almost as much about computers as she did.

Anna's Nuke War experience led to repeated lectures by her parents about cybercrime, ethics, and the consequences of misusing the power that lay at her fingertips. But nothing changed Anna's mind about Nuke Wars like being grounded from the Internet for a week at a time. Well, almost nothing.

Although Anna credits her parents and her own personal development as the reasons for her decision to abandon the IRC Nuke War scene, the real lesson came from another IRC user. One day, while "nuking people without provocation," Star Road came face to face with the nemesis of the nuker: the NukeNabber. The NukeNabber was a primitive firewall that could capture nukers in the act and show the person running the program the IP address that had attempted the offensive strike. One of the users that Anna had attempted to hit that day was running NukeNabber and responded with a message that scared her out of nuking for good.

"I'm using NukeNabber, and I know who you are," the IRC user told Anna. "I'm going to tell your ISP!"

Anna paused for a moment and thought about what this could mean. Was she in trouble? Like any daughter of somebody who worked in the legal field, she had an idea about how to find out. So she looked up the law pertaining to her online activities. That's when her young mind for the first time fully wrapped itself around the implications of the Nuke Wars. Denial of service was a computer crime. From that day on, Star Road ceased to be a combatant in the IRC Nuke Wars. She quit nuking forever and adopted a new online identity: Starla Pureheart.

^ ( ) < [ ] > ( ) *^*

Anna was 12 years old when she took the name Starla Pureheart. It was a perfect name for somebody who participated in the online Dragonlance World fantasy role-playing game. She had discovered the game through the Kender chat room. There was no particular significance to the name, however. Starla was taken from her earlier screen name, and "Pureheart just sounded good," she said.

With her lamer nuking phase in the past (as well as a short-lived "I know better than my parents" rebellious stage), Anna embarked on her true hacking education. During the next year, she discovered various "anarchy" texts that referred to the art of phone phreaking—the process of replicating the tone used by the telephone switching system to initiate free long-distance calls. The first phone phreaks, including the likes of John Draper (a.k.a. Captain Crunch) and Steve Wozniak, the famed inventor of the Apple computer, had plied their trade almost 20 years before Anna was born. Anna studied the ancient phone phreaking texts the way an archeologist might study artifacts from a lost civilization and began laying a solid foundation upon which to build her hacking education. She even attempted a few crude, unsuccessful phreaking sessions of her own, using a pay phone. Although the pay phone was conveniently located out of her parents' sight, she soon discovered that it also was located right across the street from a local police station. She was still learning.

But Anna was a quick study. One of the files she found alluded to the sheer majesty of the Linux operating system. "Learning to hack under Windows is like learning to dance in a body cast," the file stated. The author was right, as far as Anna could tell. Many of the Windows-based hacking programs she ran crashed due to faulty coding. In addition, she had managed to successfully hack her own Windows system, but soon realized that this was no big achievement. "Windows had all the built-in security of a rubber crutch," she later said.

By 13, Anna was already well on her way to mastering the Linux operating system. Windows programming and hacking had so frustrated her that she decided it was time to completely abandon the operating system in favor of Linux. That winter, her father had bought her a copy of *The No B.S. Guide to Red Hat Linux 6*, complete with the operating system on CD. Anna installed Linux on her computer, tweaked its settings, reinstalled it, and finally got it to work.

But learning and understanding Linux the way she understood Windows was no trivial task. It would take months of studying, testing, tweaking, and prodding before Anna felt comfortable in her new operating environment.

"I think of this as my first true hack," Anna recalled. "Not nuking or inept phone phreaking, but bringing Linux to life on my own machine. Needless to say, I was thrilled. I understood it, and we began dancing in C."

The C programming language was not the only dance partner that Anna had, however. The culture of the online world had become as familiar to Anna as an old friend. She was comfortable there and had made many friends. In fact, Anna has said, some of her closest friends and those with whom she shares the most in common are Internet friends whom she has never met face to face.

Psychologists and so-called hacking experts are fond of labeling such friendships as symptomatic of the estranged world and twisted personalities of the hacker underground. Hackers, they say, are introverted, antisocial, and feel more comfortable with the faceless interactions of the Internet than in traditional social settings. But the experts rely on the fringe elements of the hacker underground to form their opinions and profiles of the larger hacker community. They have all but ignored hackers like Anna Moore. And by doing so, they have perpetuated a false understanding of who hackers really are, what their true motivations are, and the contributions that these rising stars of the Internet world may make to society in general.

But the characterization of hackers as socially awkward teenagers with oversized glasses and pocket protectors no longer applies, nor does the image of the errant vandal. In the case of Anna Moore, hacking is the ruthless quest for knowledge and the study of technology. It has been one of her "most pleasurable pursuits."

Anna is also living proof that contradicts the image of the teenage hacker as dark, brooding, antisocial misfit, more likely to be caught in a drunken stupor than in the pursuit of serious hacking exploits. Of course, the hacker establishment is largely responsible for perpetuating these stereotypes; Anna, on the other hand, has never seen any attraction or usefulness in that part of the hacker scene. The beer-induced debauchery practiced by some elements and given a stamp of legitimacy by some of the so-called prophets of the hacker movement is an encumbrance to be avoided.

"The party animal mentality is incompatible with my hacker mentality," Anna wrote several months after her first DefCon experience in 2001. "What is the point of drunken carousing when I could be flying on the wings of code? You have to be sharp, perceptive, and on your toes to hack. Anything that befuddles the mind is a hindrance."

^ ( ) < [ ] > ( ) *^*

The road to DefCon started with a poster hanging on a public bulletin board at the local Buy-4-Less grocery store on Main Street in Norman, Oklahoma. A single thumbtack secured the poster to the board amid dozens of other flyers and advertisements for everything from free kittens to used cars, baby-sitting services, and garage sales. But this poster was different; it was an announcement for an upcoming meeting of the Oklahoma City–based 2600 hacker group.

Anna's mother had heard of 2600. And when Anna expressed interest in attending the meeting, her mother thought it was a great idea. It was May 2001. Anna had just turned 15. She, too, had come across information about 2600 in many of the hacking texts and Web sites she frequented. However, the 2600 Web site had listed the location of the local Oklahoma branch as a shopping mall that had long ago been bulldozed. Until she came across this latest flyer in the grocery store, Anna had been operating under the assumption that Oklahoma's 2600 group was defunct. Now she had new information, and new hope.

Anna's mother agreed to drive Anna to the meeting on the far north side of Oklahoma City, some 45 minutes away. They arrived a few minutes early and waited for the rest of the group to get there. And they waited. And then they waited some more. Either 2600 was the most secretive group in history, capable of melting into the surroundings, or nobody had showed up. It turned out to be the latter. So Anna and her mother drove home.

Disappointed, but not deterred, Anna made contact via e-mail with one of the members of the group and found out that most of the 2600 crew lived in her hometown of Norman. They were, in fact, still active, but had been delayed in getting to the last meeting because of their participation in the Norman Medieval Fair. Sometimes, cultural experiences took precedence over technical discussions about hacking.

The next attempt to hook up with 2600 was more successful. Anna arrived ready and eager to take part in unbridled technical discussions of computers, the Internet, and everything digital. She prepared herself for the proverbial hacker's data dump.

That wasn't what she received, however.

Sure, the rest of the group actually showed up this time, but the discussion left a lot to be desired. The group lacked organization. Rather than launching into a semi-structured educational session, the black-clad group of older males sat around eating greasy food-court cuisine, drinking soda pop, and shooting the breeze. Anna's mother was not pleased. To this day, she talks about the frustration of driving 45 minutes each way for nothing.

The next couple of meetings proved more fruitful for all involved, thanks largely to Anna and her mother. The two newbies to the scene produced CDs full of information about the FBI's controversial e-mail snooping program called Carnivore, backgrounders on encryption, articles and research on the alleged U.S.-led global spy network known as Echelon, and various other security issues that hackers would be interested in discussing. They made copies of the CDs and passed them out to all of the members of the 2600 chapter. A 15-year-old and her mother had brought a semblance of order and a vision to their local 2600 chapter. In return, Anna received assistance with Linux. One of the members of the group supplied her with the latest and greatest version of the Red Hat Linux operating system. This was what 2600 was all about. This was the 2600 of old, reborn.

As the meeting was coming to a close, somebody asked if anybody planned to go to DefCon that year. A few hands shot into the air, and faces lit up at the thought of the infamous annual hacker conference. Anna, on the other hand, had not yet heard of DefCon. That didn't lessen her enthusiasm, however. If it was a hacker conference, Anna wanted to be there.

"Sure, I'll go," said Anna. "What's DefCon?"

And so the journey began.

Nothing demonstrates Michele Moore's dedication to her daughter's hacking education like her agreeing to chaperone Anna and three other teenage members of the Oklahoma 2600 group (all boys) across two time zones to the City of Lights, where one of the world's most notorious underground conventions took place. It would be an education for both mother and daughter.

But getting there was easier said than done. DefCon cost money. Only one of the four teenagers had a steady job. Anna and the others, however, were able to scrounge together a few hundred dollars from their part-time consulting work designing Web pages for local businesses. Still, all had agreed that nobody was getting on the plane unless they all had cash in hand. And when the deadline arrived for purchasing airline tickets, the group was about $200 short. Anna's mother reminded the budding hackers that if they really wanted to make the trip to DefCon, it would require personal sacrifice. It was time to get out there and raise the money.

At noon the next day, Anna and her three hacker compatriots met in front of a local electronics chop shot called Mark's Computers, and the kids began to unload all of their excess computer equipment for what amounted to pennies on the dollar. Anna sold one of her extra video cards. The boys sold extra cables, modems, motherboards, and a few old system cases. When the computer gear was gone, Anna's mother drove them to several pawnshops, where they sold personal possessions in a desperate effort to raise money. Anna sold a

Pokemon game, a Backstreet Boys CD, and a Mario 64 game. The boys walked into each shop holding laundry baskets full of old and sometimes new Christmas and birthday presents, including a complete paintball set, video game systems, old sports equipment, baseball cards, and music CDs. Everything went. And by the end of the day, enough money had been raised. Anna and her friends would be going to DefCon after all.

^ ( ) < [ ] > ( ) *^*

Anna's mother had accepted a huge responsibility by agreeing to escort three teenage boys to Las Vegas. And she did not take it lightly. There were rules that the teenagers would have to follow, liability release forms that their parents would have to sign, photos and emergency contact numbers that would have to be collected, and hand-held radios that would have to be tested and issued to all four of the DefCon newbies. There was no way that Anna's mother was about to turn three teenage boys loose in Las Vegas without a tether.

The oldest boy, about 18, was very smart, conservative, and straight-laced. Anna's mother remembered him as courteous and deferential, with a quick wit and a very pleasant sense of humor. He had moderate-level hacking skills and was capable with computers. But whether or not the boy had any real hacking ambitions still remains unclear.

The middle boy was, in Michele Moore's words, a mess. He came from an unhappy, broken home and could be unruly and at times an obnoxious, loud-mouthed shock-jock. His life had been scarred by fear, and he was continually overcompensating for his sense of insecurity by maintaining a "screw-you, I don't care" attitude toward everything and everybody he came into contact with. He was continually testing the limits to see how far he could go, and often tried to test Michele's resolve and authority as the adult leader of the group. Anna's mother would have none of that.

It soon became necessary for Anna's mother to take the boy aside and "get in his face," as she said later. In her estimation, it may have been the first time in his life that an adult cared enough about him to set him straight about who he was, what his life really meant, and how his actions and words influenced and affected others. She could see from the expression on his face and the look in his eyes that it was very difficult for him to be confronted in this way, particularly by a woman. But DefCon had brought him to an emotional and behavioral crossroads. To Michele Moore's surprise, he responded with a new attitude and a renewed sense of leadership.

The youngest boy was the true genius of the group. He was a hacker of extraordinary skill, intellect, and insight. He had plans for his future and was very disciplined about daily study and practice. However, beneath a quiet, thoughtful exterior, he concealed an advanced ability to manipulate people and situations. His mind was always several steps ahead of his mouth, weighing his words carefully for effect, and silently assessing everything at all times. Anna's mother thought of him the type of hacker who has a brilliant career in front of him if he can stay clean. The potential was there—for anything.

It was to this rag-tag group of amorphous teenage identities that Michele Moore entrusted her young, bright-eyed daughter Anna. She, too, was still developing the outline of her own hacker ethic and her specific place in the hacker social order. But DefCon, held in the heart of the Vegas strip, where chintzy neon lights flicker and sparkle 24 hours a day, would be the place where she would begin to visualize the wide range of personalities and motivations that constituted the community of which she was now a part.

^ () < [ ] > () *^*

The group arrived in Las Vegas on July 12, DefCon registration day. The heat was unearthly. The concrete jungle of the strip offered little hope for shade and sucked the energy out of the unsuspecting tourists in short order. The youthful exuberance of Anna's male counterparts convinced them that walking the six blocks from their shuttle stop at the Tropicana to the site of DefCon at the Alexis Park Hotel was clearly the way to go. However, in reality, those six blocks equaled almost seven miles! Anna and her mother hailed a taxi.

There were two registration lines to choose from: one for White Hats and another for Black Hats. Anna and her mother stood in the White Hat line. They looked around, watched, and listened as they collected their registration materials. In many ways, DefCon is responsible for its own reputation and for the preconceived notions that first-time conference attendees bring to the show. Anna and her mother arrived with their own set of preconceived notions and opinions about what they were about to experience. One look around the hotel confirmed most of their expectations.

A cash bar had been set up in the center of the large Parthenon ballroom and was already doing a great deal of business. In fact, rumor has it that the cash bar alone generates enough money to pay for twice the cost of the entire range of hotel repairs that are needed every year after the conference. Other conference areas were still being set up. The hotel was in a state of semi-controlled disarray. This was the kind of weekend that Anna and her mother had prepared themselves for.

Anna and her mother sat down at a table to organize the dozens of papers, flyers, and conference agenda forms. They studied the full-color program, taking in all of the bright graphics and illustrations and the strange, ominous language used to describe each of the technical presentations. It went on for eight pages. There were presentations scheduled on how to write back-door hacker programs and create remote buffer overflows and "So you got your lame ass sued: A legal narrative." Anna and her mother skimmed the conference program from front to back. Then they turned to each other slowly, their eyes met, and they burst into uncontrollable laughter. DefCon would be everything they had imagined it to be—"a combination of machismo, sexism, science fiction, and bathroom humor," according to Anna's mother.

Anna simply shook her head, laughing. "Oh brother," she said. "Here we go."

Despite its toned-down demeanor compared to previous years, DefCon remained true to form. On the first night, roving bands of drunken troubadours serenaded the hallways of the Alexis Park into the wee hours of the morning with broken renditions of various rock songs. The next day, those same crooners could be seen catching catnaps on the floor of the ballroom between sessions.

During the second night, the Drunken Whores hacker club hosted a social-engineering contest in the large air-conditioned tent set up on the roof of the Alexis Park Hotel. But when one of the contest organizers suggested a prank that would be a clear violation of federal law, things quickly got out of control. One of the feds who was present in the audience informed the individuals that if they did what they were proposing, he would have to arrest them. At this point, Anna, her mother, and the boys made an orderly exit. A few seconds after they managed to get away, somebody in the crowd yelled that the feds were about to arrest everybody. It was a scene of utter panic, as hackers of all shades began a chaotic rush toward the exits. This was a far cry from the Oklahoma 2600 meetings.

On the third night, an ambulance arrived with sirens blaring. Emergency personnel whisked off to the hospital a DefCon attendee who had fallen victim to a serious case of alcohol poisoning. This was the type of weekend it had turned out to be. But there was a bright spot in all of this mayhem. Anna had won the CyberEthical Surfivor contest.

^ ( ) < [ ] > ( ) *^*

When Anna was proclaimed the winner of CyberEthical Surfivor, her mother embraced her, and a crowd of 2600 colleagues gathered around. Dozens of hackers offered congratulations, handshakes, minidisks, business cards, and contact information. Reporters requested interviews and photographs. In addition to an $800 donation of books on computer ethics that was made to her local library in her name and a coveted DefCon leather jacket, Anna won a lifetime free pass to DefCon. It was the best reward of all, she said. "And I certainly intend to use it."

However, the contest ended with a strange sense of disappointment lurking just beneath the surface. The judges couldn't hide their dissatisfaction with the contestants, who were not as unethical as they had expected them to be. It was DefCon's first foray into ethics, and hardly anybody had expected it to be taken seriously.

As Anna stood on the stage to accept her victory, Richard Thieme, the third judge in the contest and a self-proclaimed Internet visionary, made a few closing remarks that could only be described as a desperate attempt to revive the image of irreverence and debauchery that most knowledgeable observers of the hacker scene now agree has been one of the hacker community's biggest problems. "Some of us have been shocked and a little grieved to the degree to which the cyberethics game really did focus on ethics," said Thieme. "Where is the old hacking ethos? Where is the spunk, that vitality that drives you across boundaries no matter what, to get what you need, in order to have your parents notified when you're less than 16 that you have done it? We think she [Anna] has the spunk of a real hacker in her heart."

It was a confusing message from somebody who claimed to be a leading thinker in the hacker community, and a sad way to congratulate a young hacker for taking the issues seriously. But the ethos that Thieme spoke of was not Anna's, and for good reason. Up and coming hackers like Anna don't fit into that twisted world of old, where ethical behavior is defined by a blind devotion to the idea that something that can be done should be done. Anna's spunk and vitality as a hacker are as strong as the motivations that powered the first generation of hacking pioneers. The moral and ethical relativism that Thieme hinted at in his closing remarks have no place in Anna's definition of what a real hacker is.

When it was all over, DefCon left a lasting impression on the 15-year-old girl. It was a world of contradictions. Far removed from the drunken depravity and the dark alleyways of the electronic underground, she found an atmosphere of complete trust among the gathering of more than 5,000 hackers. A stranger with a long white, scraggly beard asked Anna and her mother to keep an eye on his laptop computer while he went to a seminar. He also allowed Anna to use the system to send e-mail to her father in Oklahoma.

The DefCon "goons," as the security personnel are known, put Anna at ease, making her feel safe. DefCon was the ultimate hacker party, as far as Anna was concerned. From technical seminars to computer games to rave dancing and hacker mischief and education, it was everything she had expected, and more.

DefCon, however, was only the beginning. Anna is now taking high school courses through the Independent Study program at the University of Oklahoma. Last December, the Independent Studies department began a concurrent high school/college program, and Anna started earning high school and college credits simultaneously. She is the first student to participate in the concurrent program.

When she is not trying to improve her practical hacking knowledge, Anna is focused on preparing for her black belt exam in tae kwon do. She's trained for four years and intends to continue developing her skills and abilities for the rest of her life.

Hacking, however, remains a passion. But it is a passion that Anna feels she can navigate safely thanks to the moral and ethical compass that her parents have helped her develop. "As I was crossing into greater areas of knowledge, my parents wanted to be sure I could self-regulate and make wise decisions," Anna said, looking back on her regular discussions with her mom and dad about hacking. "They caused me to understand that the freedom they gave me came with a price—responsibility." Anna feels confident that she's learned that lesson well.

Could this be the beginning of a new age in hacking? Could hacking one day be viewed as a positive influence on teenagers? The answer to those questions is undoubtedly yes. The face of hacking is changing for the better. And a young girl in Norman, Oklahoma, is helping to lead the charge.

7

Unlikely White Hat: Willie Gonzalez

Today is the day his life will change forever. From this point on, he will question everything: parents, school, friends, religion, right and wrong. His way of thinking will change. Things will never again be black or white. No longer will there be only one way to do something. There will be endless shades of gray, possibilities where once there were none. Curiosity will become his mistress, a close friend, a warm blanket on a lonely, chilly night.

For a 12-year-old kid in Rogersville, Tennessee, this is the kind of change that comes along once in a lifetime. It's a small town, nestled between Knoxville and Kingsport in the northeast corner of the state, where everybody knows everybody else. Change happens in Rogersville, but it rarely sneaks up on people the way it is about to sneak up on Willie Gonzalez.

It's just another hot summer day in Tennessee, and Willie is answering telephones with the teenage son of his father's boss. Willie's father, a heavy drinker who's been divorced from Willie's mother for six years, drives a delivery truck for a small company headquartered in a one-room office in a small, dilapidated garage. Willie spends his days there passing the time. There are two computers in the office, each located on a desk not far from the other. The owner's son is typing something on one of the keyboards and seems to be in a trance as he stares at the screen.

Today the owner's son is typing faster, more frantically than usual. The small office is filled with the sound of tapping on the keyboard. The taps are loud and come in quick, intermittent spurts, like a jackhammer. The low rumble of office chair wheels rolling across the plastic mat on the floor, back and forth between the two computers, punctuates the lulls in the typing. Then a few more rapid strikes of the keys, and the chair wheels rumble across the floor again. The process is repeated.

Willie doesn't know what to make of this. So he walks over to where the owner's son is sitting and stares over his shoulder.

"What is that stuff?" Willie asks, looking at a screen full of cryptic numbers and letters and a small digital timer in the lower-right corner that is ticking down by the second. "What does it say?"

"This is a terminal emulation program," the teenager says, never once turning his head away from the screen. "I'm breaking into a remote agency's computer system."

Willie stands there for a moment trying to process what he has just been told. Then he looks around the office and toward the door to see if anybody else heard what he just heard. Nobody is there. What the hell is the owner's son talking about? What is a remote agency?

Without warning, the teenager swings his chair around and pushes himself across the floor in one swift motion to the other system and begins typing. Then he flips nervously through a spiral-bound notebook, mumbles something to himself in a foreign technical language that Willie does not understand, and begins typing again.

Willie can feel the heat of the blood moving through his ears and his cheeks. The sound of the keys clicking on the computer keyboard is raising his blood pressure. It strikes a chord deep within his soul, the way the sounds of airports, train stations, and cruise ships make some people want to travel to the far ends of the earth. It's contagious.

"Would you like to help?" the boy asks Willie, catching him off-guard.

"Sure. I'm pretty good with computers. I used to have one." Willie is referring to an old Commodore 64 that his grandmother had bought for him four years earlier. It was a crude system by today's standards, but it had been powerful enough to ignite Willie's interest in computers.

The boy instructs Willie to sit down in front of the other system and, on his command, to hit the Enter key. Willie does as he is told. He nervously prepositions his finger above the Enter key, not sure exactly what is happening but sure that he doesn't want to screw it up.

"Hit it!"

Willie slams his index finger down onto the key. In an instant, with the press of one key on the keyboard, text begins to scroll up uncontrollably on the computer screen. It's clear to Willie that he's no longer sitting in a ramshackle old garage. With the press of that key, he's entered a new world. It was that easy. When the text finally stops scrolling, he reads the words on the screen:

```
WARNING! You have reached a United States Gov-
ernment computer system. Unauthorized access is
prohibited by Public Law 99-474 (The Computer
Fraud and Abuse Act of 1986) and can result in
administrative, disciplinary or criminal pro-
ceeding.
```

Willie focuses on the important words, especially that first one, written in capital letters, and the last two, which seem ominous. He's not sure if what he's reading is real or part of some game that the owner's son is playing. Regardless, it's exhilarating, like something out of a James Bond movie.

Then the owner's son swings his chair around beside Willie and pushes him out of the way. The teenager begins to type. He types faster than Willie has ever seen him type before. With a grin on his face and a sparkle in his eye, he turns to Willie and says, "Nice job. I'm in."

Willie sits back and watches as more text fills both screens. A few minutes later, it is all over. The owner's teenage son relaxes his posture, lets out a loud sigh of relief, and closes his eyes.

"What just happened?" Willie asks the boy. "What did you do?"

After a few seconds of silence, the teenager begins to tell Willie about the power of computers; what they are capable of, what people can make them do, and how they work. Then he pauses, turns his chair toward Willie, and says that he's what people call a hacker.

"Do you know what a hacker is?" he asks Willie.

Willie shakes his head. "Not really."

The teenager then describes for Willie what a hacker is, what hackers do, and why they do it. There's no talk of destroying data or stealing sensitive information. The satisfaction and the thrill come from the hack, the teenage boy tells Willie. It is enough for him to be able to get in, he says. Defeating a system that others proclaim to be secure is like discovering the ultimate truth.

Soon the discussion draws to a close, but it will not be the last. This is only the first of what will be a series of semi-educational lessons about hacking that will take place in that small garage office. It also is the beginning of the rest of Willie Gonzalez's life—a life as a hacker.

^ ( ) < [ ] > ( ) *^*

That summer of hacking was more than a life-changing experience for Willie Gonzalez; it was a rebirth.

Born in Miami, Florida, in January of 1976, Willie was the younger of two sons raised by a mother of Cherokee Indian decent with few professional skills and a Cuban-born father who could hold his liquor better than he could a job. The family moved often between Florida and Tennessee. Somehow they survived.

Willie remembered a home life full of uncertainty and fear. Virulent shouting matches between Willie's mother and father erupted often and without warning. His parents fought about everything, from his father's addictions to alcohol and cigarettes, to the lack of money to pay the rent. Willie's father was famous for disappearing after these fights for weeks at a time. Still, Willie's mother managed to keep food on the table. Whether it was borrowed from a neighbor or bought with the little money his mother managed to stash away, Willie and his brother always had something to eat.

When their parents divorced, Willie and his brother were faced with the difficult choice of deciding which parent they wanted to live with. Although Willie's father was a drunk, he maintained a strange power of persuasion over the boys. The boys decided somewhat reluctantly to leave Florida and live with their father in Tennessee. It was a decision that Willie would later regret.

Life in Tennessee was more difficult than anything Willie had experienced before. His father's drinking steadily grew worse, as did his ability to hold down a job. When the money ran out, so did the heat for the run-down wooden shack that Willie and his brother called home. Winters in Tennessee could be brutal, especially at night. The boys wore more clothes to bed than they wore to school. They wrapped themselves in sweaters, wool caps, and jackets—anything they could find.

Welfare checks paid for the food, but delivery was left to the only reliable method of transportation the family had: Willie's two feet. Even during the winter, when the driving wind blew the falling snow horizontal to the ground, Willie trudged several miles each way to the grocery store to pick up the family's food. Then it would be up to him to prepare the meal and clean up afterward. If Willie didn't do it, it didn't get done.

Making friends didn't come easily for Willie. What people in Rogersville didn't understand, they disliked. And they didn't understand the dark-skinned Hispanic kid named Willie Gonzalez. In fact, for a long time, maybe a year or longer, Willie didn't have a friend in the world. He didn't fit in. He rarely spoke. He ate lunch alone and never volunteered to answer questions in class. He internalized the hurt and embarrassment that came with always being chosen last for sports and games. His grades suffered.

For a while, Willie turned to sports as an escape. He had always been a fast runner. Soon he was breaking records as a halfback for the junior high football team. Things were looking up. Suddenly, people wanted to know him, and they judged him based on his performance on the field, not on the color of his skin, his ethnic background, or his family's wealth. For the first time in his life, he knew what it felt like to belong to something. Or so he thought.

Willie's sports career ended with a single tackle at the opposing team's goal line. He remembers how it felt when his knee buckled under the weight of half a dozen defensive linemen. The pain was unbearable. The crushing blow tore a ligament and fractured his left leg below the knee. The resulting surgery not only ended his hopes for a future in sports, but it also ended the sense of belonging that Willie had worked so hard to achieve. Soon he discovered that his newfound acceptance in school went only as far as the football field. He was back again at square one. His grades plummeted to where they had been during his loneliest of times.

Willie was heading down a dangerous path that a lot of teenage hackers and wannabe hackers find themselves on. He was an outsider, a loner with no escape from a dreadful home life. For a long time, the only person who took the time to talk to Willie in a meaningful way was the teenage son of his father's boss—a hacker. And he, too, was somewhat of an introvert and a loner with antisocial tendencies. But he helped raise Willie's awareness to the wider world around him. He showed Willie that although computers can be ruthlessly demanding of their users, they never prejudge the person who sits down in front of them. If you know what you are doing, the computer accepts you and cooperates without question. And best of all, computers can take you away from places you don't want to be. They are gateways to other worlds. And Willie was looking to escape.

^ ( ) < [ ] > ( ) *^*

Willie was 14 when he decided to move back to Miami and live with his mother. His father's drinking had gotten out of control, and high school life in xenophobic rural Tennessee had become unbearable.

The move forced Willie to rediscover his relationship with his mother, who had done very well for herself since divorcing Willie's father eight years earlier. She had remarried and now owned her own produce company. The change of scenery for Willie was drastic. Compared to Tennessee, life in Miami was heavenly. His mother kept a clean, country-style home, and everyone, including his new stepfather, shared the daily chores. More important, life in Miami afforded Willie the financial freedom to feed his obsession with computers.

Willie's mother never had a difficult time figuring out where her son was during these early years. If he wasn't at school—soon to be the rule as much as the exception—he more than likely was at one of the many computer hardware warehouses located on 72nd Avenue between 25th and 36th streets. This was the area better known as the Silicon Valley of Miami. If you needed something for a computer, there was a store somewhere along this stretch of Miami pavement that sold it.

The young hacker built his first real computer by hand. It was only a 386 running ancient versions of Windows and DOS, but to Willie it was a modem-equipped work of art that sat in his mother's spacious living room. It was on this system that Willie learned about hacking. He "owned" his system from the start. He learned how to optimize the system's memory, how to use a DOS shell and configure every aspect of the machine.

Willie's Internet surfing started shortly after his stepfather took him on a software-buying spree. All of the games and applications that he purchased offered online bulletin boards where users could go to download updates, patches, and software drivers. Willie soon discovered thousands of these bulletin board systems, or BBSs as they are called. And many of these BBSs offered pirated copies of software applications that sold in the stores for hundreds of dollars.

One day Willie embarked on a reconnaissance mission of sorts. He discovered files that belonged to other BBS users and ways to gain access to and change the settings of certain BBSs, as well as various user account management tools. But his most valuable discovery was the online presence of other kids his age from Braddock Senior High School in southwest Miami who also were into the BBS and Internet Relay Chat, or IRC, scene. Almost immediately, the rag-tag group of cyberpunks began to meet at school to trade BBS account information and access numbers the way other kids traded baseball cards.

Willie became entrenched in this brave new world of like-minded individuals. Once again, he felt like he belonged to something bigger than himself, only this time it wasn't some official organization led by uncaring adults. In the online world, your age doesn't dictate your social standing. And that was important to Willie and his IRC friends. After all, they were hackers, and hackers control their own destinies.

^ ( ) < [ ] > ( ) *^*

It wasn't long before computers and hacking became more important to Willie than school. To Willie, high school in Miami sucked. Many of the teachers didn't care about the students or the quality of their lessons. Willie talked back to his teachers and routinely arrived at his classes late and unprepared. He complained that the school didn't give him enough time to fight his way through the 5,000 other students who crowded the hallways.

Sure, it was easier being a Hispanic kid in Miami than in Tennessee, but school still sucked. Willie simply wanted high school to be over with. It was an inconvenient part of daily life, and he tried to miss as much of it as possible.

On most days, Willie skipped school by himself. He spent a lot of time walking up and down Miami's Silicon Valley and often ducked into a store called Incredible Universe, a gigantic, air-conditioned warehouse that sold all of the latest electronic gadgets and computers. Willie explored and experimented for hours with the floor-model computers, pretending that his parents were shopping somewhere in the store.

In the aisles of Incredible Universe, Willie learned the ins and outs of all of the newest applications and operating systems, and he had a blast pulling electronic pranks on unsuspecting customers. He got a kick out of watching customers react with fear to an error message that popped up telling them they just broke the computer. Even better was watching some shopper follow a series of commands that caused the hard drive to be reformatted, and then the look on the salesman's face as he tried to figure out how every dim-witted customer managed to reformat the hard drive.

Even on the days that he skipped classes, though, Willie eventually made his way to school. But by 3:15 in the afternoon, he usually was home, and by 3:30 he would be linking up to a local ISP. Within 10 minutes of connecting to the Net, Willie would be logged in to half a dozen systems around the country, chatting with his newfound cohorts on IRC and planning the rest of the evening's activities. He would take a short break for dinner and then return to his room for the rest of the night.

Willie adjusted his mind and his environment to accommodate his change in work and sleep habits. The truth was he was getting very little sleep. He painted the walls of his room a dark blue-gray, installed a black light in his lamp, and hung a blue shade on his window that matched the blue in the carpet. On the walls, he hung pictures of wiccan—a religious denomination based in ancient witchcraft—neopagans, and Egyptian gods. The black light transformed the pictures into psychedelic living organisms that moved to the rhythm of the electronic "trance" music that played from a tape in Willie's stereo. Now when he surfed the darkened corridors of the Internet, he was in the correct frame of mind. The computer was a mere extension of his body, and he an extension of it.

It took awhile for Willie to get warmed up. His serious journeys into cyberspace rarely started before 2 A.M. This lead to a lot of problems between Willie and his mother, who was becoming increasingly upset about his waking up late for school. She conducted the equivalent of an FBI raid on Willie's bedroom at least three times a week, threatening each time to take away the computer if he didn't shut it down at a reasonable hour. That ignited several furious exchanges between mother and son. Willie equated his mother's threats to take away his computer with being threatened with jail and isolation from all contact with the outside world. It was a punishment he could not and would not live with.

The computer had become a way of life for Willie, the sole means of communication with most of his friends and the gateway to his comfort zone. He wasn't about to give it up cold turkey. In a desperate attempt to keep his mother off his back, he tapped on the keyboard as tenderly as he could and even resorted to locking his bedroom door to provide a tripwire to alert him to his mother's late-night raids. He would do whatever he had to do to keep his mother at bay, even if that meant doing well in school for a while. What else could he do? After all, there was a war going on. And Willie, who was now known online as Spirit, was one of the main combatants.

^ ( ) < [ ] > ( ) *^*

Bewm! The digital bomb ripped through Willie's system, spewing a deadly barrage of zeros and ones that his computer's Windows 95 operating system didn't know how to defend against. The bomb severed his Internet connection, stopping the flow of life-giving data to his system. The screen turned a pale blue color, as if its blood flow had been cut off. It was 3 A.M., and Willie was wounded. He had to reboot, redial his ISP, and reestablish his connection to the IRC server he had been using to chat with friends. He had just been nuked.

Unprovoked as it was, the attack had an immediate impact on Willie and one of his IRC buddies from school. It sent them to the library looking for answers and new ways to defend against the attacks. It also turned them into IRC vigilantes in search of the culprits.

The "blue screen of death," as it was known, was the result of the now infamous denial-of-service attack tool called WinNuke. It targeted systems running Windows 95 and Windows NT by sending a packet of what is known as out-of-band, or OOB, data to port 139 on the target host. Unfortunately, when a system accepted this packet, the result was a near instantaneous crash that left the user cut off from the Internet and staring at a screen full of error messages. In the early days of the Nuke Wars, the payload of choice for the attacker was a packet containing the word *bewm*, as in "Boom! You've just been hit with a blue screen bomb."

Willie's hacking education went into high gear as the IRC Nuke Wars began to heat up. He hooked up with a few friends at school, one of whom was in possession of at least a hundred root accounts at various corporate and university systems around the world from which the group could launch their nuker counterstrikes. This kid with the root accounts specialized in creating high-speed, coordinated denial-of-service—DoS—attacks. He was a good kid to have working for you, especially when an attack called for the targeting of routers at the enemy's ISP.

Willie specialized in surveillance and reconnaissance—both skills that would come in handy later in life, though he didn't yet know it. Willie's job right then was to collect data on the enemy without being caught. He prowled the Internet like an invisible spirit—hence his hacker nickname. Spirit was sharp, quick, and quiet. His calling card was ominous: "Careful, that cold chill traveling down your spine may be me looking over your shoulder," he told people online. He put together dossiers on his targets: country of origin, presence of firewalls, type of OS in use, vulnerable ports that are open to certain types of protocols, connection speed, ISP, and any other details he could collect. Spirit also specialized in spoofing his own Internet identity. He maintained dozens of reference pages in a binder that listed proxy servers: domains and e-mail servers that he could use as relays to make it more difficult for others to track his movements.

IRC had become the Wild West of the Internet. Traipsing through IRC was like walking alone through the streets of Fort Apache in the Bronx after midnight: you needed protection. Point-and-clickers—the horde of unskilled script kiddie hackers—were everywhere. To defend himself, Willie relied on cracked firewall "warez."

Going on the offensive in the IRC Nuke Wars was easy. It truly was a point-and-click attack. But developing an effective defense against being nuked took a little more knowledge. Sure, Microsoft had issued software patches to fix the hole in Windows that made it vulnerable to nuke programs, but new versions of the attack software hit the Net in a matter of hours, compared to the weeks and months it took Microsoft to develop and issue new patches. Willie had to learn a lot on his own.

He started by figuring out how to configure rule sets on different types of firewalls. WinNuke attacks also forced him to become familiar with protocols, like Internet Control Message Protocol, or ICMP. Computer networks use ICMP messages to troubleshoot problems, such as when a router is unable to transmit data packets as fast as it receives them. ICMP messages communicate these problems between systems automatically.

Things got out of control when full-scale war broke out. No chat room or multiplayer Internet game was safe from being hijacked by an IRC script kiddie armed with a nuker program. Willie and his friends didn't appreciate it when some "lamer" who was about to lose some game decided to nuke the other players. They were equally pissed off at the posers who enjoyed hijacking their IRC channels, taking over operator status and booting them out of the chat rooms.

Soon Willie and his friends from high school formed a loose-knit group of IRC warriors. They worked in tandem to identify and harass specific individuals on IRC whom they knew were responsible for widespread WinNuke attacks.

On the weekends, Willie routinely logged 18- to 20-hour days sitting in front of the computer, and at least 10- to 12-hour days during the week. The tactics being employed in the IRC Nuke Wars were becoming more serious and demanded more time and energy. When it wasn't feasible to identify a specific machine belonging to an IRC nuker, the next target automatically became that person's ISP. Willie and the other combatants referred to the process as "holding down an ISP." It involved launching a DoS attack against the routers of the ISP that served the accounts belonging to your IRC nuker enemies. ISP attacks almost always succeeded in helping you regain control of a chat room and opening up an enemy's accounts to direct DoS attacks.

The era of the IRC wars didn't last long, but its impact on Willie's view of security was profound. The ability to gather information and intelligence has always been an important tool in the security professional's toolbox. It's also a critical skill for hackers. And Willie made use of it now, and he would again, five years down the road.

^ () < [ ] > () *^*

Willie hated the typing class that his school required him to take during his sophomore year. He couldn't type with his hands in the so-called "proper" position. It was unnatural. Willie was a hunt-and-peck typist. His hands needed to be able to move freely across the keyboard, and he needed to look down at his fingers every now and then.

Frustrated by his unwillingness to adjust, Willie's teacher gave him exactly what she thought he'd been asking for: a failing grade. She put him in a seat in front of a computer located at the back of the class and tried to forget about him. He wasn't worth her time; he didn't want to pass. She was wrong, of course. Willie wanted to pass. He just didn't want to play by the teacher's rules. His new seat at the back of the class was exactly what he'd been hoping for.

Left to his own devices, Willie quickly figured out how the Word Perfect files that were used to grade the students' typing assignments were stored on the network. To his surprise, they were all saved in the same directory. All Willie needed to do now was crack the teacher's network account.

The next day, he came up with a plan and enlisted the help of one of his IRC buddies, a fellow veteran of the nuker wars. They hatched an elaborate scheme that called for Willie's friend to create a diversion long enough for Willie to lace the teacher's keyboard with chalk dust. Then, when she logged on to her computer, a second diversion would be set in motion to enable Willie to inspect her keyboard for clues to her password.

As soon as the teacher walked in, Willie's friend jumped up from his chair and yelled that his computer had a virus. "It's deleting all of my work," he said, throwing his hands in the air. It was actually a script that he and Willie had written that did nothing but print virus-like warning messages on the screen, but to the untrained hacker eye, it looked like it was really deleting files.

The teacher didn't even bother to put down the stack of books and papers she was carrying. She just rushed back to investigate the ballooning crisis. That was Willie's cue. As the rest of the class gathered around his friend's computer, Willie quickly walked up to the front of the class and pretended to sharpen a pencil. He grabbed a pre-positioned eraser that was filled with chalk dust and silently sprinkled the chalk over the keyboard. Then he walked back to his friend's desk and nodded: phase one was complete. That was his friend's cue to hit the Escape key, killing the script, and then pretending that all was back to normal.

Willie's friend watched closely as the teacher typed her password and hit the Enter key on her computer. Then he initiated phase two of the operation.

"It's happening again!" he shouted. "This time I think it really is deleting files. And they look like files that belong to the rest of the class."

The teacher, Mrs. "I don't know that I'm being scammed," ran back to investigate again. Willie pretended to help by looking at everybody's screen, checking for signs of a wider virus infection. One of the systems he checked was the teacher's. But he wasn't looking for a virus. He was taking note of the letters on the keyboard that had no chalk dust on them. Those were the letters that made up the network password belonging to Mrs. "I don't know why there's chalk all over my keyboard."

The plan worked like a charm. In fact, it was almost too easy. Willie returned to his computer at the back of the classroom and began to guess the password. It was eight letters long, which Willie thought could be a problem—that is, until he realized that the letters spelled p-a-s-s-w-o-r-d.

From that point on, Willie had the run of the school network through a privileged account. He mapped the entire network. It wasn't long before he was surfing through all of his classmates' grades.

When the next graded assignment rolled around, Willie waited until one of the better students in the class had finished her work, showed it to the teacher, and received her grade. Then, once the girl's file was closed, Willie sprang into action from his computer. He changed the name at the top of the file to his name, raised his hand to indicate that he was done, and to his teacher's astonishment turned in a flawless assignment. Suddenly, it didn't matter how Willie positioned his hands on the keyboard. He was passing.

With his hack of the school network complete, Willie realized that he was standing at a digital crossroads. Should he turn left and leave his mark on the system, wreak havoc on the network, and seek to instill fear in the hearts and minds of the teachers he despised so much? Or should he slip back into the shadows?

^ () < [] > () *^*

Early one Saturday morning, Willie was standing on the second-floor balcony overlooking the living room of his mother's house. In front of his old computer sat a young boy, not yet a teenager, who went by the online moniker Kryp. His mother was a close friend of Willie's mother, but this was the first time that the boy had visited with her.

The young kid was trying out Willie's new X-Wing space combat simulation game, and he wasn't doing very well. For a while, Willie stood there watching and chuckling as the boy grew increasingly frustrated with the computer game. Then Willie introduced himself as Spirit and offered to teach Kryp the finer points of X-Wing, including the cheat codes and the natural limitations that the game inherited from its human coders. For a young kid, Kryp was quiet, calm, and focused. Most kids his age can't sit still; but he was self-possessed. The questions that he asked stretched Willie's knowledge to the limits. The boy's mind was two steps ahead of the computer at all times. His hands moved across the keyboard like those of a virtuoso playing a finely crafted stringed instrument. Willie liked the kid's attitude and was impressed by his commitment to learning and understanding the logic of the game and the way in which the computer processed the code. That's how hackers play computer games, Willie thought to himself. A few hours later, the boy walked out of Willie's house carrying a copy of the game and various other software applications that Willie had saved for him on floppy disks. The student had met the teacher.

The first thing Kryp did when he got home later that evening was call Willie on the telephone so that the older, more experienced gamer could walk him through the setup of X-Wing. The installation process was pretty straightforward, even for a young kid like Kryp. But when the game didn't play, he needed Willie to walk him through a series of tweaks to his system to boost its performance. Willie taught Kryp how to manipulate the memory on his system and how to run various DOS utilities and adjust configuration settings. Again, Kryp asked all the right questions, and even a few that Willie couldn't answer. But what really impressed Willie was that once he instructed Kryp in how to do something, he never had to repeat himself. The conversation lasted several hours. By midnight, Kryp was flying through space playing X-Wing.

Kryp was mesmerized by his first hack. Willie had taken him on a journey through his computer, giving him his first taste of what it's like to make a system do what you want it to. From this point on, Spirit and Kryp were no longer just friends; they were mentor and apprentice, professional associates, members of an exclusive club of the like-minded few. They were fellow hackers.

Within a year, Kryp was calling on Spirit's expertise regularly. Since that first hacking lesson on the telephone, the young boy's skills and interest had progressed rapidly. He liked the idea of being a hacker. And his appetite for hacking had taken him online and into the realm of security. He wanted to know why things work they way they do. All of his questions now revolved in some way, shape, or form around security and access protections to games.

Willie never gave him a straight answer. He treated Kryp's hacking questions the same way he treated his initial gaming questions: he offered definitions and basic background information, but allowed Kryp to connect the dots on his own and forced him to ask more questions and explore. One of his recent areas of study was FTP, or File Transfer Protocol. FTP is the protocol that is used by computers to send and exchange files via the Internet. Willie walked Kryp through a connection to a local FTP server that stored thousands of files and programs, including hacker utilities. He explained the commands and the host's responses. He showed Kryp how to pull up a help screen and how to list a directory of files, but Kryp was already there. That was a lesson they had covered already, and Kryp had learned it well. This was all very basic for Kryp. He was more interested in the content, the individual files and programs that are stored on the FTP server. Willie could see that Kryp knew what he was doing, and that what he was really fishing for was help finding FTP servers that offer free hacker utilities. But Willie wouldn't go for it. He refused to simply hand him information. Kryp would have to work for it and learn on his own.

"You know how to use FTP. Now go out there and find out what FTP really offers and how you can use it," Willie told him.

Thus ended another simple lesson. The student had homework to do.

^ ( ) < [ ] > ( ) *^*

By the time Willie was 18, his thoughts began to shift toward the future. What was he going to do with his life? Was the level of hacking he had engaged in so far all there was for him? His mother had been thinking and asking the same questions. She had walked in on him in his bedroom a number of times when he was allegedly playing games on his computer. But as the media's obsession with hackers began to reach a frenzied state, Willie's mother began to worry about her son.

"What exactly are you doing?" she asked. "You aren't doing any of that hacking stuff, are you? Is the FBI going to come knocking on my door because of it?"

She had a point. There was good reason for her to be nervous. Willie was living the *rebel* in *rebellious*, skipping school, talking back to teachers, and barely graduating. And she knew that he'd been trying for years to conceal his late-night Internet activity.

At first, Willie skirted the issue. But hacker incidents began making the news in record numbers. Willie's mother, who had never even flipped the power switch on a computer, grew more concerned, especially after word got out that a 19-year-old hacker had been deemed so cunning and dangerous that a judge had ordered him not to even talk about computers.

Willie tried to explain hacking to his mother. Hacking is the exploration of technology and of the self, not an exercise in destruction, he told her. "Hacking is a way of life," he explained. "You don't just hack a computer, you hack your car or your school assignment, too. Hacking is an unconventional way of thinking and solving a problem that cannot be solved by conventional means. You see, mom, hacking is not specifically related to computers."

But for Willie, hacking was more about computers than anything else. "Computers are an important part of my life," he told his mother. "It started out as a curiosity, then grew into an interest, then into a hobby, and now a passion."

Willie's mother sat in silence, listening to her son talk about hacking like it was a religion. He actually used the phrase "way of life," and that really threw her for a loop. What was to stop Willie from ending up like the other kids she was reading about in the newspapers?

The truth was that Willie had all of the makings of a serious black-hat hacker. He was a frustrated, rebellious, eager-to-learn teenager with a little bit of skill. There was nothing to stop Willie from breaking the law except Willie.

^ ( ) < [ ] > ( ) *^*

Today Willie's going to do a series of things that on any other day and at any other company would get him sent directly to prison. He's going to walk right into the new, massive call center of one of the biggest cellular telephone companies in Miami, pretend to be somebody he's not, get an ID badge, introduce himself to everybody as "the new guy" on staff, find an empty computer cubicle, do a little social engineering, and run the L0phtCrack password cracking program to gain root access to the network.

The year is 1996. The FBI has reported that more than half of all cyberattacks against companies originate from internal company systems. More than one-third of the attacks come from the Internet, according to the FBI. News has also surfaced that a group of unidentified hackers recently defaced the Web site of the Department of Justice, replacing the site's contents with pornographic images, swastikas, and the label "Department of Injustice." Hackers also hit the U.S. Air Force's Web site, shortly after a PC that stored hundreds of thousands of credit card numbers mysteriously walked out of a Visa International office. And Kevin Mitnick has just recently pleaded guilty in Los Angeles to federal charges of cellular phone fraud.

Willie is 19 and fresh out of a job with a local Miami company where for the past year he'd done quality-control inspections, tests, and research on computer equipment. That hadn't been the most exciting job in the world, but it paid the bills. The only thing that really kept him there was his access to computers and the slow, but steady, recognition he received from the corporate brass for his skills working with the network. He had eventually moved into network and user troubleshooting. Then one day he was asked by the CEO to surreptitiously recover, by any technical means possible, sensitive corporate data that one of the company's salesmen was suspected of pilfering. It was a welcome respite from the daily grind. Willie succeeded and thought it was his big break as a hacker. In the end, all he got out of it was a handshake and a free lunch with the CEO.

Today will be different, however. The tedium of quality control and user support feels like a lifetime ago. Today, he's going to raise the ante.

It was late, well past the morning rush hour, when Willie walked into the building through the opulent lobby and onto a waiting elevator. He pushed the button that would take him to the second floor. He was dressed to kill: long-sleeve shirt with fancy cuff links and a buttoned-down collar, a sleek-looking paisley tie, pleated dress pants, and shiny, black dress shoes that looked like they'd been hand-crafted in Italy. He looked like he belonged on the elevator. Based solely on his appearance and demeanor, Willie was above suspicion.

When the doors of the elevator opened, Willie stepped out and turned toward the receptionist with the confidence that said, "I know where I'm going. I've done this before." In fact, he had done this before, late, after hours.

"Good morning. I'm Joey Simpson," Willie said with a smile. "I'm supposed to start work today in the call center. Mr. Jones told me to get a temporary access badge from you until my permanent badge can be made."

"Of course," the young receptionist responded, returning the smile eagerly. "Just print the badge number and sign your name in the log. The call center floor is through those double doors behind you." She smiled again. "Welcome."

"Thanks. Do you need me to return this badge to you at the end of the day?"

"Yes. If you don't have your permanent badge by tomorrow, we'll just issue you another temporary one."

"Okay. Great. Have a nice day."

It was that simple. Willie never even flinched. His voice had remained steady and confident, and his body language had supported perfectly his claim that he was a new employee, still wet behind the ears, in search of a little help and compassion from a co-worker. It worked every time, he thought as he walked away confidently.

Willie slid his guest badge through the card reader next to the door. It clicked, and a small green light blinked, giving him the go-ahead signal to enter. Then he jerked open one of the heavy double doors and stepped onto the large call center operations floor. The door closed behind him, making a loud *whoompf* sound. Willie took a moment to scan the sea of cubicles, looking for one that hadn't been claimed by another member of the burgeoning workforce. There was an open cubicle in the corner, where he would be able to see people approaching him. It was perfect.

As confidently as he had walked out of the elevator, Willie made his way through the call center. He made a point of introducing himself as Joey

Simpson, the new guy, to every employee who took notice of him. They were all thrilled about Willie's—Joey Simpson's—decision to join their ranks.

Willie grabbed a few manuals and laid them on the desk beside the computer, opened, of course. This made for a convincing show. The computer was already running. The only thing Willie needed was a network login ID and password. So he put on a show of confusion, the type of confusion that eager first-day employees often display when they are trying to figure out their new computing environment. He waved the floor supervisor over to his cubicle.

"Hi, Mrs. Williams. I'm Joey Simpson." Willie extended his hand. "Today's my first day, and I'm having a little trouble getting into the network."

"Hi. It's nice to meet you, Joey. Actually, I wasn't aware that we had anybody starting today," the supervisor responded.

"Oh. Maybe Mr. Jones forgot to mention it. I still have a temporary badge until they make me official." Willie stood there, confident, but wide-eyed and eager. This was truly one of his Academy Award–winning performances.

"Well, don't be insulted by my ignorance," Mrs. Williams said. "We're expanding so rapidly that there are new faces that I don't recognize popping up around here every day. I'll log you in using my account for now. Then just give the IS department a call to make sure your account gets set up as soon as possible."

"Okay. Great. Thanks very much. I appreciate your help."

"No problem. It's good to have you aboard," Mrs. Williams said, smiling. "Let me know if you need anything else during the course of the day. I know how crazy first days around here can be."

"I will. Thanks."

As soon as Mrs. Williams disappeared into the sea of work cubicles, Willie sat down in front of the computer and did a quick survey of the applications. It was a Windows NT system. He opened the database application just in case somebody walked by and surveyed his work area. Then he reached into his left shirt pocket and removed a couple of floppy disks. One of the floppy disks contained a program called L0phtCrack that would help him uncover weak user passwords on the network by looking for passwords that match common words in the dictionary. Another contained a file called SAMDUMP.EXE.

The first thing Willie looked for was the SAM file in the Windows NT registry. The SAM, or Security Account Manager, file is a repository of all user information. However, the file, which was stored in a directory called \winnt\system32\config\sam, was locked. Willie had expected this; the file was being used by the operating system. But quickly he struck gold. He located a backup copy of the file, known as the SAM repair file, in the directory

\winnt\repair\sam. To his surprise, the \winnt\repair\ directory was shared, meaning that anybody who knew what they were looking for, including a hacker like Willie, had access to it.

Willie copied the SAM repair file to a phony directory that he created and ran the SAMDUMP program against it. That expanded the SAM file and put it in a format that would allow him to apply the password cracker utility to it. Then he inserted the floppy disk containing the L0phtCrack program and ran the password cracker against the SAM file. He thumbed through his call center manuals while the program did its magic.

L0PhtCrack first runs a common dictionary attack, looking for passwords that match words found in the dictionary. These are the easiest passwords to find and are cracked very quickly. Then it moves on to a brute-force attack, using thousands of combinations of letters, numbers, and special characters. In a matter of seconds, L0phtCrack breaks the simplest passwords—the ones that are three or four letters long. In less than 15 minutes, Willie had dozens of six-letter passwords. In 20 minutes, he had more than a hundred, including the network administrator account. In hacker terminology, Willie owned the network, like a god.

The adrenaline was pumping through Willie's body with the force of water shooting out of a fire hose. His senses were operating at maximum capacity. This was the biggest hack he'd ever pulled off. But then the consequences of what he'd been able to do began to sink in. This was too easy. The livelihoods of 500 people were in his hands. He could easily use the access he'd gained to scramble the brains of the entire company. Operations would come to a screeching halt, customers would go ignored, and people with real lives, real families, and real bills to pay might lose their jobs. Willie thought to himself: Aren't they lucky that I'm the one who pulled this off, and not some unscrupulous hacker.

As nonchalantly as he had entered the call center, Willie downloaded the hundred or so user IDs and passwords, placed the floppy disks back in his shirt pocket, got up from his chair, and walked out. He wasn't carrying a brief case or wearing a coat, so he didn't look like he was leaving. But he was. And soon he was heading toward the company's main office, its corporate headquarters.

At the main office, Willie repeated the steps he'd taken at the call center: he punched his floor number into the elevator, walked confidently through the halls, and greeted the receptionist. But here things are different. He's no longer pretending to belong here. He does belong here. He has a desk, a telephone, and a computer that are assigned to him, Willie Gonzalez, not some imposter named Joey Simpson. The call center hack he'd performed only moments earlier, as elegant and daring as it was, wasn't a crime; it was part of a security audit ordered by the company's senior executives. And Willie is a network security specialist. It's his job to hack the network.

^ ( ) < [ ] > ( ) *^*

Willie was barely old enough to buy a beer when a career in information security came knocking. This was too good to be true. There were companies in the world that were willing to pay him to work on computers, to hack, as long as he agreed to abide by a certain set of rules: ethics.

But that wasn't a big stretch for the kid who grew up a lonely outsider. Although such early life experiences can sometimes turn good kids into demons bent on nonconformity at any price, Willie's childhood in Tennessee taught him compassion and humility. While some kids who experience the type of hardship and rejection that Willie experienced choose to lash out, others look for an escape, a distraction. Willie looked for an escape, and he found one in computers and hacking. It became his way of reaching out to a world that had rarely shown a willingness to accept him. Although the physical world was full of narrow-minded, judgmental adolescents, the hacker underground did not judge you based on your ethnicity or financial status. In the online world, the bottom line was what you could do with a computer.

"I always saw computers as a way to open myself to others and share that piece of me with them," Willie recalled years later. "And I didn't want that piece of me that I shared to be destructive, or constantly bent on destroying what others had created."

Willie's newfound maturity didn't happen overnight. No single event shaped his views of hacking or hackers. It was a gradual progression. More important, his hacker ethics came to be defined by his experiences in the real world, not the fantasy world of the Internet. Real people and real events helped to shape his sense of right and wrong. Sure, he'd had his fun taking part in the IRC hacker wars and had witnessed a hack of a government system before he even knew what hacking was, but it was an adult desire to not be like the kids that had rejected him in Tennessee that had opened his mind enough to see the good in people. Staying on the right side of the law, therefore, and displaying an undying respect for technology became Willie's way of just saying no to the circumstances of his youth. To the mature Willie Gonzalez, going down the road to criminal hackerdom would have been equivalent to admitting defeat.

Willie had become a hacker in the truest sense of the word. He loved the technology. "You cannot honestly love something and at the same time try to use it to harm others, or destroy something that someone has made," Willie said. "If you truly love something, you explore it, you respect it, you learn to understand it and use it in ways that benefit others."

That is, as far as Willie is concerned, the hacker's ultimate responsibility. Hacking is about the search for truth, the love of technology, and, more important, sharing the wealth. And what Willie also came to realize was that when a hacker shares that wealth of knowledge, he does so at the risk of changing another person's life. Being another hacker's mentor comes with an entirely different set of responsibilities, especially when that hacker is only 11 years old.

Willie stopped using everything related to the name Spirit: instant messaging accounts, ISP accounts, e-mail accounts, and IRC chat rooms. He started a new life with a clean slate and a new sense of the person he had become.

"Taking on the responsibility of teaching someone about hacking really made me look at it in a whole new perspective," Willie recalled. "It just wasn't me anymore. I was a role model for somebody to follow whether I liked it or not. Sometimes you don't know how well you understand something until you have to explain it to someone else."

Willie reevaluated his beliefs and his reasons for hacking. He interrogated his inner hacker, searching for answers to the questions that had been plaguing him for so long, and that had only recently begun to plague his best friend Kryp. Why was he interested in hacking? What were his goals? What was it that he was trying to accomplish through hacking? Who was it that he really wanted to become? How did he want people to think about him? How did he want to be remembered, and for what?

The answers to these questions don't come easy, nor do they come all at once. It takes time and soul searching.

★ ^ ★ ( ) < [ ] > ( ) ★ ^ ★

Willie and Kryp were sitting in a local Taco Bell in Miami, inhaling a dozen thin, hard-shell tacos and sipping sodas that looked too large for one human being to finish. Taco Bell had become their classroom, similar to the mall food courts used by the dozens of 2600 chapters around the country. They'd been meeting here at least once a week for the past three years to talk about hacking. They called it going on a "DP run"—a Dr. Pepper run.

Since that first lesson on how to hack his system to install X-Wing, Kryp had become like a younger brother to Willie. DP runs served many different purposes and were not limited to conducting hacking data dumps. The conversation usually started on the topic of girls and gradually progressed to school and various other personal problems. But most of all, they talked about hacking.

"I've been hacking into some systems on the Net," said Kryp.

"What systems?" Willie asked, alarmed but not really surprised by the revelation. "You're not breaking into government systems are you? You'll get busted. You're not ready for that."

"No. They're just some lame sites that nobody knows exist anyway. Plus I'm not doing anything to the servers. I'm just hacking my way in and leaving. I'm doing it mainly for practice and to learn," Kryp replied.

"So, you're not stealing anything or obliterating anybody's servers, right?"

"No way. You taught me better than that," Kryp said. "Plus, real hackers don't get into that anyway. It's not about destroying systems."

"Are you covering your tracks?"

"No. I'm leaving my name and address. Duh? Of course I'm covering my tracks. The log files are history, man. They've got nothing on me. If they did, or wanted to do something about it, they would have done something by now anyway."

By this time, Willie's appetite had dissipated. He'd gone from shoveling tacos into his mouth with savage force to simply staring at his food. His best friend had just confided to him that he was hacking into systems on the Internet. Willie wasn't prepared for this. He had clearly underestimated Kryp's abilities, not to mention the influence that the hacker scene on the Internet was having on the young kid. This must be what parents go through as they watch their kids grow up, Willie thought.

Suddenly, Willie's inner hacker began to talk to him. He felt responsible for Kryp's future. He knew from their previous conversations that Kryp's parents knew nothing of their son's hacking interests. It was up to Willie to steer Kryp in the right direction. After all, he had helped launch Kryp's hacking career in the first place.

"Why do you want to hack?" Willie asked.

"I don't know. Because I can. It's cool." Kryp eyes betrayed a sense of unease and even hinted at a trace of embarrassment at what he soon realized was an uninspired answer.

"C'mon, Kryp. Give me a break," said Willie, shaking his head in disapproval.

"Okay, okay, okay. Hacking is cool, but I love beating the computer at its own game. It's an awesome feeling to be able to make computers do what you want them to do. Plus the system administrators that run these sites are completely lame. They don't know what they're doing. They can't possibly feel the same way as I do about the technology."

"So you love the technology. Is that what you're saying?"

"Yeah," said Kryp. "But I want to ask you something."

"Shoot."

"Am I a hacker?"

The question straightened Willie's spine.

"Are you a hacker?" Willie repeated the question with an incredulous tone in his voice. "Well, what do you think? Do you think you're a hacker?"

"Yeah. I've gained root access to more than a hundred Web servers, and I own at least two-dozen .edu domains."

This was the moment of truth. Whatever Willie chose to say at this point and how he chose to say it would forever shape Kryp's outlook on hacking. The young boy wanted desperately to be a hacker. Willie knew that. But Kryp was like most kids his age; they wanted to be hackers without knowing what a hacker really was. Most kids weren't as lucky as Kryp; instead of hooking up with a mentor like Willie, the vast majority of teens allow themselves to be duped by the less scrupulous members of the underground. Willie chose his words carefully.

"Kryp, you are a hacker," said Willie. "But you're not a hacker for the reasons you just spelled out to me. You're a hacker because you learned programming languages, mastered countless operating systems, protocols, and programs. You learned how to manipulate computer systems to work for you, to achieve what you needed them to achieve, even when they were designed not to. You expanded your mind beyond what was written in books, or what was explained to you. You allowed your knowledge to grow and never limited yourself or what you could learn. You not only hacked computer systems, you hacked your life itself. You hacked yourself in self-discovery, and you grew beyond the boundaries that others set for you. But you need to understand that there is no classification system for hackers. Being able to break into a system doesn't make you a hacker."

The look on Kryp's face told Willie that he had made contact with his friend's inner hacker. Kryp shook his head up and down in agreement with what Willie had just said. His arms were folded, and he stared out the window and watched the cars drive by.

"Do you know what the difference is between a hacker and a criminal?" asked Willie.

"What?"

"Respect. A hacker has respect and a keen understanding of technology. A criminal has no respect for society's laws or for others and merely sees technology as a tool to use in exploiting the system to get what they want."

"So are you calling me a criminal?" Kryp asked.

"No. I'm trying to make a point. For example, a kid who cracks into a Web server, defaces the main page, and replaces the text with his own text and images is no better than a common teenage delinquent who breaks into somebody's car to steal CDs, or who spray-paints graffiti on a public billboard. Yet they are quick to claim the title of hacker. Somehow they feel that calling themselves hackers justifies what they did. What they really are is a bunch of common criminals."

"Yeah. You're right. But I don't deface any sites. That's lame. I'm not hurting anybody or destroying any information."

"You don't know that for sure," said Willie, growing impatient with Kryp's reluctance to comprehend the ramifications of his actions. "I know what it is like to seek out more challenging networks and pit yourself against their security systems. I would finish hacking one system and then it was on to another one, with tougher security to crack. An unfortunate truth to all of this is that when we're young, we sometimes don't realize the repercussions of our actions. We don't fully comprehend the reality of supporting a family, or how our actions can ripple through the lives of others. You've got to accept that responsibility if you want to call yourself a hacker and if you want to train yourself on systems that don't belong to you."

"Okay. I get it now. I understand," Kryp responded, looking Willie directly in the eyes. "I don't have to break the law or hurt people to call myself a hacker—that's what you're saying."

"Exactly," said Willie. "Just remember that. I don't want our next DP run to be in the county jail."

Kryp laughed, which triggered a reluctant smile from Willie. This wasn't the first time Willie had had to explain the essence of a hacker to Kryp. Kryp had been asking Willie the same questions for years. But Willie had been hoping he'd be able to find the answers on his own.

"Let's get out of here," Willie said.

^()<[]>()*^*

Willie went on dozens of DP runs with Kryp during the following year. The questions were always the same, as were Willie's answers. Then a new job presented itself and forced Willie to move away from Miami. Somewhat reluctantly, and with a slight sense of trepidation, Willie left for the small town of Rockledge, halfway up the Atlantic coast of Florida, on the shore of the Indian River Lagoon.

And now Willie and Kryp still talk on the telephone, but things aren't like they used to be. Now Kryp is busy going on DP runs with a new crop of teenage hackers, helping them to find their own places in the world. When they ask him technical questions, he doesn't give them the answers. He forces them to believe in themselves and to find solutions on their own. That's how Willie taught him.

Invariably, the question is asked, "Am I a hacker?" Kryp tells his younger friends about an old hacker he used to know. He imparts the lessons that Spirit taught him years earlier.

"Hacking is neither good nor bad," he says. "It's people that are either good or bad."

And although the choice is up to them, Kryp tells the young hackers that the most important lesson he's ever learned is that breaking into a system is not what makes you a hacker. "Spirit told me that," he says. "And he's the unlikeliest white-hat hacker I've ever met."

8

Tinker, Teenager, Hacker, Spy: The H.D. Moore Story

This was the day that everything Rick Fleming thought he knew about hackers got turned upside down. As the technical director of computer security operations for Computer Sciences Corporation's San Antonio, Texas, office, Fleming had seen his share of talented hackers. After all, the multibillion-dollar consulting firm was deeply involved in the Air Force's intelligence and information warfare program run out of nearby Lackland Air Force Base. Fleming, a veteran Air Force intelligence officer, had worked on a number of computer security and intelligence programs throughout the years, most of which remain concealed behind a veil of military secrecy. During those years, he had also earned a master's degree in computer and network engineering. He knew what he was doing. And he liked to think that he knew a hacker when he saw one. But that was about to change.

It was March 1999. Fleming's people for the past two months had been telling him that it would be worth his time to check out some new hotshot hacker from Austin named H.D. Moore. It seemed like everybody knew about this kid, or at least his hacking prowess. He's a real expert programmer, they were saying, and already has a solid reputation among military types for being one of the brightest new minds in hacking and security. If Fleming were to believe everything he was hearing, then there was only one conclusion to reach: H.D. Moore was one of the best hackers in the world, and Fleming needed to hire him before somebody else did.

Fleming was always on the lookout for new talent, especially the kind of talent that came along once in a lifetime. If H.D. Moore was as good as Fleming's own people were saying he was, then he was probably worth the time it would take to interview him. But even today, the day of the interview, Fleming had his doubts. H.D. Moore was young, very young. In fact, he was still in high school. That's right; the kid who was about to arrive for an interview to work on classified military information warfare programs was only 17.

When H.D. pulled up in front of CSC's offices after an hour-and-a-half drive from Austin to San Antonio in his broken-down Audi, Fleming was surprised at what he saw. The kid was wearing a polo shirt and dark trousers. His hair was cut short and combed neatly. It wasn't exactly what Fleming had expected from a teenage hacker. And little did he realize that for H.D., the yuppie look was a strange adjustment from his usual black leather jacket and Mohawk.

Of course, Fleming hadn't approached H.D. Somebody familiar with Fleming and the CSC operation had done so months earlier in, of all places, an Internet Relay Chat room—the IRC, that hangout of hackers, where conversations took place in real time. Fleming had no idea what was said to the young hacker to gauge his interest or that the IRC contact had come right out and asked him if he wanted to work for Air Force intelligence. That wasn't Fleming's style. But for now, it worked. The kid was in his office.

Fleming told H.D. to take a seat and offered him something to drink. It was obvious that H.D. was nervous. When Fleming questioned him, H.D. reacted as if he were being judged because of his age—and, to a certain extent, he was right. But Fleming had a 16-year-old son and felt comfortable in his knowledge of the teenage intellect. H.D.'s answers came shooting out of his mouth in a barrage of low, muffled sounds. He had the unique ability to speak as quickly as his mind processed his thoughts. Fleming strained to listen.

Before grilling H.D. on his specific knowledge, Fleming decided to break the ice by asking about H.D.'s background and experience. H.D. said he was about to graduate from an alternative education program in Austin that allowed him to make his own hours, take the courses that he wanted to take, and earn credit for various computer projects he was working on outside of school.

Then, as unusual as it might be for a 17-year-old, H.D. presented Fleming with references from some of the biggest companies on the Internet. He told Fleming that a year or two earlier he had discovered a bunch of serious security problems on the Web sites belonging to Yahoo! and Microsoft. "I contacted their system administrators and told them what I had found, and they agreed to provide a reference whenever I applied for a job." Fleming was genuinely impressed. And there was more.

"I also wrote a program called NLog and worked with the Navy's Shadow project," H.D. said.

"You worked with the Shadow guys?" Fleming leaned back in his chair. Not only was H.D. only 17 years old, but now he was telling Fleming that he had professional experience working for the Navy's Shadow security team. Shadow—otherwise known as Secondary Heuristic Analysis Systems for Defensive Online Warfare, an intrusion-detection program, sort of a burglar alarm for computer networks—was launched by the Navy to help weed out subtle probes and attacks against military networks that standard commercial firewall systems can't detect.

"Yeah. I took the current code base and rewrote large chunks of it so it could be deployed in a single computer configuration," H.D. responded. "I gave a presentation about NLog, too. The SANS Institute asked me to do that. NLog automatically logs scans conducted against network ports. Then it dumps all of the scans into a database and then you can do detailed analysis to uncover subtle scans. That's good because most people don't know that they're being scanned."

Fleming was fascinated. "Tell me more about the Shadow software. What do you think about it? What's unique about it?" Fleming knew the answers to his own questions, of course. Now he was testing the depth of H.D.'s understanding of networks and his grasp of the subject matter.

"It's different from the normal pattern-matching systems currently available because it doesn't care about what is actually in the packet, just what the packet looks like and how it relates to other packets," H.D. continued, trying desperately to impress Fleming with his depth of knowledge. "The term for this is traffic-analysis versus signature-matching intrusion detection. And with Nmap, you have a situation that allows hackers to map entire networks, even the entire Internet, and begin to plan for future attacks based on the vulnerabilities they find." Then he concluded: "I think we're going to see more coordinated attacks in the future."

"So you think these scans should be taken seriously, as a threat to national security, is that correct?"

"Yes. In less than 10 minutes, every machine that responds to an ICMP echo request or ping can be mapped and its operating system fingerprinted. These scans should be taken seriously and should be considered a warning of future attacks."

Fleming, who'd interviewed hundreds of security professionals during the past two years at CSC, couldn't believe what he was hearing. The young kid sitting in his office had been working on the Shadow program and spoke about future threats like a sophisticated professional. In fact, a few months earlier, the Pentagon had detected signs of highly coordinated scans against military networks that seemed to originate from multiple locations all over the world. It was then that the Shadow team at the Naval Surface Warfare Center had sent an official alert to the rest of the defense community that an internationally coordinated reconnaissance was underway using the Nmap tool. The possibilities were endless. But the most serious implication was that some group or nation was systematically mapping the U.S. military and commercial Internet infrastructure, locating vulnerable machines, and saving that information for a possible future information warfare attack.

The threat was real and completely plausible. H.D.'s analysis of Nmap was exactly correct. The tool was virtually untraceable. In its most sophisticated form, it sends one or two packets of data from different locations into a stream of millions of friendly packets heading for specific servers. One or two packets of data can slip right under the radar of most intrusion-detection systems because most administrators set the threshold of their intrusion-detection alarm systems at a much higher level. Nmap takes all of the brute force out of scanning and reconnaissance and, therefore, makes stealth attacks much easier to pull off.

In September 1998, when Navy analysts first discovered the Nmap scans, they uncovered a coordinated effort that spanned 15 different locations on several continents. The attackers sent as few as two probes per hour. During the week of September 13, scans were observed coming from 14 different Internet addresses, including major ISPs. Five different attacks scanned a single firewall within minutes of each other. Some attackers sent as few as two packets per day. The scans continued for three days and stopped within a few hours of each other. Officials were convinced that it was a long-term, coordinated effort to map the location, configuration settings, and vulnerabilities of important military and civilian servers.

But H.D. was as adept at understanding the defensive mechanisms inherent within Nmap as he was at understanding the tool's attack capabilities. That was, after all, a large part of why he was sitting in Fleming's office right then. CSC and the Air Force were interested in his ability to help them secure their networks as much as they were in his programming skills and the potential application of those skills in the world of offensive information warfare.

"Determining the source address of an Nmap scan isn't as hard as the general opinion makes it out to be," explained H.D. "Even with decoy packets, you can use the process of elimination to narrow down your attacker's real IP address: context clues, previous traffic from one of the source addresses, and the more aggressive techniques like determining which, if any, of the source addresses are capable of launching Nmap, for example. The most common source addresses I have seen in Nmap scans have been from dial-up setups, university systems, and the occasional compromised ISP server."

"Tell me how you went about figuring out how to trace the Nmap scans back to their source addresses," Fleming asked, clearly intrigued by the teenager's grasp of complex networking theory.

"The original method I used to track the source addresses of Nmap scans involved randomizing TTL fields"—Time to Live fields, which specify how many hops a packet of data is allowed to make before finding its correct destination on the Internet. "What I did was I looked at the TTL fields for all of incoming packets from an Nmap scan. Then I took each of the source addresses and sent one ping to get the to/from TTL pairs for that host. After you determine the number of hops a packet takes from the source address to the scanned machine, you can start guessing original TTL values for the Nmap scan, and you can see which hosts fit the profile. By far the fastest way I have found to detect an Nmap scan and a variety of spoofing attacks is by using a Perl script that hashes packets based on their sequence numbers and then prints a report if the same sequence number occurs in a number of packets above a threshold."

Fleming had heard enough. After only an hour of talking to H.D., Fleming knew that if there was a way to hire the kid, he'd do it. But most companies, including CSC, don't have a mechanism for hiring underage teenagers as apprentices. As a result, hackers like H.D. were an untapped talent pool. H.D. had an idea, though. He said he would incorporate himself. Then he wouldn't just be a kid; he'd be a legal corporate entity. Fleming grew even more impressed. This proved that H.D. was a real thinker. He was not your typical teenager. There was more to him behind the quiet demeanor and fast talking. There was a lot more.

^ () < [] > () *^*

Born in 1981, H.D. lived in Hawaii until he was six. That's when his parents split up and sent his world into a tailspin. After his parents' divorce, H.D.'s life became something of a circus act. He was forced to move back and forth from town to town and start over again in school after school. His memories were of always being the new kid and living with whichever parent was less screwed up at the time. Fighting became part of each school day. He was kicked out of more than a few elementary schools for fighting, including one where he didn't last a full week.

When he was 10, he moved in with his mother in Austin and began to finish his last two years of elementary school in an upscale part of town. Despite his surroundings, life became even more difficult for H.D. His mother was poor, and his clothes were threadbare. The other kids picked on him incessantly, especially about his name. He refused to tell people his real name. It was something that was out of his control anyway. It was just his bad luck to be born at a time when his parents were going through some new religious fad and decided it would be cool to give their son a Hindu religious name. Again, there were fights. But as far as H.D. was concerned, his name was H.D.

H.D. buried himself in books, reading tons of science fiction and fantasy novels. He also loved sitting in front of the computers at school. He would arrive at school early in the morning before the rest of the kids and tinker with the Apple IIe that was in the library. It was quiet there, and for a short while he was lost in another world. It probably helped a little, too, that the first computer he got to play with was one that was designed by a hacker named Steve Wozniak.

For the next two years, H.D. was a ship without an anchor. During his fifth-grade year, he took a class in the fundamentals of BASIC programming. It wasn't long before he was writing his own programs. He programmed the school computers to draw strange lines on the screen in random patterns. As it turned out, H.D. was actually writing what would become widely known as a screen saver program. To him, it was just one way to make the computer do interesting things, and occasionally crash.

Seventh grade saw H.D. on the move again, this time to Florida to live with his father and attend a brand-new school. His stepmother owned a broken-down 486 PC and allowed H.D. to exercise his curiosity on it. Eventually, he got it to work. One day, he borrowed a game from a friend and discovered his first computer virus. It was then that he realized how vulnerable and fragile computers are. The virus cut through his system like a jagged blade, severing all of the vital software links that kept the system running. It was a real eye-opener for H.D., who was now beginning to move beyond BASIC programming. However, H.D.'s time in the sun in Florida, along with his father's new marriage, was short lived. Within a year, H.D. was back in Austin with his mother.

Back in Austin, H.D. made a friend who shared his intense interest in computers. And H.D.'s new friend had already taken his interest to the next level: dialing into bulletin board systems on the Internet. H.D. slept over at his new friend's house from time to time, and the two of them devised ways to hijack users' accounts by pretending to be system administrators. They succeeded, but never did anything with their newfound skills.

On the other hand, the two of them constantly got into trouble for less virtual transgressions, like taking part in vicious fights between rival groups of teenagers. Gang fighting was a real problem in certain regions of Austin. H.D. played the role: leather jacket adorned with silver biker spikes, spiked hair, a lean and muscular physique, and lots of attitude. He took part in some of the fighting as well. Eventually, half a dozen members of a rival gang got busted for stabbing and killing another kid. It was serious business, serious enough that the Austin Police Department formed a gang suppression unit. And it was a problem that officials at McCallum High School wanted to avoid if they could.

Within a year, the school notified H.D.'s mother that her son was being kicked out for "not living in the district." He had to attend a different school, the school officials said, because the rules are very specific about zoning regulations. In reality, the school was looking for an excuse to get rid of H.D. He was a bad seed, part of the crowd of troublemakers who the school was trying, futilely, to weed from its ranks.

H.D.'s new school, located on the east side of town, proved to be an even less constructive environment for a kid with a rebellious streak. Now H.D. was a minority white kid in a mostly black student body. The fighting increased in both frequency and intensity. Blows were exchanged mostly as a result of callous taunting from other students about the way H.D. dressed.

Before long, officials at H.D.'s new school began searching for a way to expel him. They treated him like they treated all of the kids who needed some guidance and stability in their lives: they told him that he was more trouble than he was worth, and that it wasn't a matter of if he'd be expelled, but when. H.D. didn't care, though; he was about to move anyway.

Before the school officials could act on their threats to expel H.D., his mother moved the family to the south side of Austin, and H.D. changed schools yet again. That made half a dozen schools in 15 years, and he wasn't done yet. A different set of changes were afoot, though. H.D. was about to bury himself in computers and hacking. It would be his saving grace, his escape.

^ ( ) < [ ] > ( ) *^*

It all started with a slow 486 computer and an account with America Online. H.D.'s mother now worked from home as a medical transcription specialist for a local hospital, and to do her job, she received a new computer, three phone lines, and a high-speed ISDN line running from the house to the hospital. H.D. got the use of the family's old 486 and began surfing the Net on AOL.

One of the first things H.D. did was get his hands on a copy of a piece of software known as ToneLoc. The tool, known officially as Tone Locator, is the classic DOS- and Windows-based war dialer. Hackers use it to dial telephone numbers automatically and save all of the connections that are made with modems in other computers. H.D. spent his first few months of systematic hacking making dozens, maybe hundreds, of random calls all over Austin looking for systems to connect to. He found some very interesting, and some odd, systems just sitting out there on the Internet, including heating, ventilation, and air conditioning systems; industrial equipment; and many banks and large corporations.

From war dialing, it was a small step to more serious programming, especially for a kid like H.D., who has a voracious appetite for knowledge about computers as well as a seemingly innate ability to understand how they work. Through his Web surfing and various conversations on IRC, H.D. came across a pirated copy of Visual Basic, an application development tool sold by Microsoft. H.D.'s mind absorbed the intricacies of programming in Visual Basic with ease. In a matter of months, he was programming tools to automate the mass mailing and distribution of pirated software.

It was during the waning months of 1997 that H.D. discovered that the software used by millions of AOL subscribers to connect to the Internet was easy to abuse with custom programs written in Visual Basic. He'd kicked his research up a notch and was delving into the way the AOL software references its internal system resources via Web addresses. Then he wrote a program to scan ID numbers and log the titles of every window that pops up when a user enters a specific address, known as a Universal Resource Locator, or URL. But H.D. had done more than simply log Web page titles and URL addresses. He'd discovered multiple back doors leading directly to AOL's back-end administration systems as well as to most of the so-called private areas of the service.

Most hackers H.D.'s age would have stopped there and either abandoned their pursuit or used their knowledge and access to do bad, if not illegal, things. Not H.D. He wanted to take his research a step further and learn how the entire AOL protocol worked. He poured through countless binary files in the AOL software. It would be a sure way to go blind for the average user, but it was casual reading for a serious hacker like H.D. After hours of reading hundreds, if not thousands, of lines of code, he found one file that contained the host name of the actual AOL server. Then he found out which port it used to connect to the Web.

One night, he changed the host value of the AOL server to *localhost* followed by what is called a null byte. The null byte, \0, is typically used to tell the computer that a data string has terminated. However, what H.D. soon discovered was that by using a null byte within an HTTP (Web) query, he could alter the URL input parameters, such as directory paths, file names, and command access settings. Therefore, applications that do not check for null bytes when performing validity checks on user input can be fooled into exposing files, operating system information, and command-line access.

Next, H.D. wrote a program that listened to port 5190, the port used by the AOL server to connect to the Net. He connected to the real AOL server and injected the false information. This allowed him to analyze the actual protocol and, if he wanted, intercept or modify messages passing from the client computer to the AOL server. With just a little bit of common sense and deductive reasoning, H.D. was able to inject his own packets into the data stream. There it was: AOL and its users were sitting ducks for skilled hackers. H.D. Moore, a teenager from Austin, had proven how simple it was for somebody without any formal training in programming or network engineering to crack into one of the biggest computer networks in the world.

^ ( ) < [ ] > ( ) *^*

At about the same time that H.D. was busy discovering serious vulnerabilities in AOL, the Nuke Wars began ravaging IRC channels all over the world. Script kiddies were employing a software tool called WinNuke to knock each other offline and take control of chat rooms. Chances are if you were on IRC during this time period, you experienced what it was like to be nuked with the "blue screen of death."

But the Nuke Wars also gave rise to a virtual arms race in cyberspace. H.D. became one of the principal arms dealers, supplying the so-called packet warriors of the Internet with DOS-based scripts that he wrote himself. He attached different hacker handles to each script, so nobody knew for sure who had written the attack code. To H.D., writing attack code or scripts that sought out Internet users who unwittingly set up their systems to share hard drives and files was legitimate scientific research. He was a hacker. He pushed Internet security technology to its limits. It was what he did. How somebody chose to use one of his scripts was a matter of personal ethics. He didn't need to use his tools to break the law or steal information. That would have been too easy, a cop out. Writing the code and deciphering the security vulnerabilities was the hard part.

In the real world, getting along with his mother also proved to be difficult. H.D.'s mother had remarried since divorcing his father 11 years earlier. However, after only four months, H.D.'s new stepfather was dead from a nagging illness. The blow was devastating to his mother's psyche, and the stress played out in her relationship with H.D.

He was on the move. Again.

H.D. arrived in California in the middle of the 1997 school year to live with his father. Along with the change of scenery came another high school, another round of being the new kid, another round of run-ins with bullies, another round of teachers who reacted to what they perceived to be an attitude instead of tapping into his energy, and another round of trying to make new friends. This time, however, H.D. put a stop to it. Within seven months, he was back again in Austin with his mother. He was tired of starting over, tired of what he called the "idiocy" of his high school experience. There was just no reason to keep trying. H.D. dropped out, a casualty of the system.

^ ( ) < [ ] > ( ) *^*

The first thing H.D. did after dropping out of high school was log on to the Internet. It wasn't that he hadn't been logged on before, but now he really put his mind to it. He logged on in a spiritual sense and dedicated real time to his hacking education. And time he had.

It was during the summer of 1998 when H.D. started work on what would become one of his first major hacking programs. In a few short months, he coded a complete remote command shell utility using Visual Basic. What he didn't know, and what few in the hacker community recognized, was that the tool H.D. was creating was, in fact, one of the first fully functional 32-bit Windows network Trojan horse programs—software that surreptitiously installs itself on target systems and sits in hiding, allowing a hacker to come and go at will.

H.D. made a few copies of the program and gave it to other hackers he knew from IRC. Unknown to him, however, some or all of those hackers took his program, broke into major "warez" sites—Web sites that offer free downloading of pirated software—and replaced the executable files of many of the most popular software titles with his Trojan horse software. Suddenly, everybody who connected to those Web sites and downloaded what they thought was a pirated version of a commercial software application was actually installing a back door into their systems. Once installed, the program sent the attacker an e-mail message with detailed instructions on how to connect to the infected system. Ironically, the Windows systems belonging to some of the biggest uber-hackers of the day were compromised, and all of their shell accounts were stolen.

H.D. was truly ahead of his time. The program that would become the most notorious Trojan horse, Back Orifice, was still months away from being released. However, when the infamous hacker group known as Cult of the Dead Cow released the first version of Back Orifice (a play on the title of Microsoft's Back Office software) in August 1998, portions of the code were inspired by vulnerabilities in the Microsoft operating system code that H.D. had discovered.

Gradually, H.D.'s coding and hacking skills outgrew Windows. He was too good, and Windows was too weak, unstable, and easy to compromise. Finally, he made the switch to Linux. Then he dropped Visual Basic and began to teach himself C programming and various other scripting languages. What amazed him most was the number of powerful and effective attack tools available for Linux. He was in scripting heaven.

One of the most amazing tools he came across was the Nmap port scanner. He couldn't believe how easy the program made it for him to determine the type of operating system running on a target host. He simply needed to send a few packets of data to the system, and Nmap would return a list of the services that were running, including all of the details on hundreds of operating system configurations. It also offered a simple way to scan large numbers of hosts simultaneously and send decoy packets to help conceal the user's true origin on the Internet. And best of all, Nmap was a free shareware application.

Recognizing the sheer power of Nmap, H.D. immediately began exploring the Internet with it. He collected volumes of information from random systems. But he soon realized that he needed a way to organize and analyze all of the results he was receiving from the Nmap scans. The amount of data he was collecting was overwhelming. Manual analysis was out of the question, especially for an expert coder like H.D.

H.D.'s solution to the information glut was a series of Perl scripts that created a flat-file database from the Nmap scans. He called his invention NLog. The CGI scripts that he included in the package allowed him to search scan logs for hosts that matched specific criteria, such as open ports, port states, operating systems, and network configurations. And being a child of the Internet, H.D. Web-enabled his application. With NLog, all of the annoying tasks involved in analyzing Nmap scans, such as dumping remote procedure call, or RPC, services and looking for NetBIOS and vulnerable Network File System shares, could be performed with just a click through a convenient Web browser interface. It became what H.D. called "the network browser of choice for lazy hackers." The tool was simple, but effective, a hallmark of its creator.

^ ( ) < [ ] > ( ) *^*

When the school year started again in September 1998, H.D. decided a high school diploma might be worth one more try. He understood the level of hacking skill he'd acquired, and not going back to school would probably kill any chance of working with computers as an adult. College wasn't necessary, but a high school diploma was, and H.D. knew it.

H.D. contacted McCallum High School for information on how to enroll for the new academic year. But officials at McCallum made it clear that he wasn't welcome. Once again, they trotted out something about zoning regulations, saying that H.D. didn't live in the school district. In fact, they would have had to allow H.D. to return to school if he had pushed it. But he didn't. Officials at McCallum recommended that he try an alternative school called Gonzalo Garza Independence High School.

Gonzalo Garza was different, and so were the students. The majority of the 300 or so students who attended with H.D. were ex-gang members. Some were hardened criminals who took leaves of absence to defend their homes from rival gang members. Almost all of the girls were pregnant, or at least it seemed that way to H.D. And despite the school's stated mission to "choose peace over conflict," more than a dozen of the students were killed before the year was over. But at Garza, it didn't matter what you did in the past—only what you did in the present and what you planned to do with your future.

The new school was, in H.D.'s words, "awesome." He made his own hours, took only the classes that he wanted to take, completed his assignments according to his own schedule, and even received credit for the computer projects he worked on outside of school. H.D. used his NLog program as one of those projects. On March 2, 1999, after H.D. contacted a few members of the Navy Shadow project about his new program, the SANS Institute, a professional association of thousands of system administrators located in Bethesda, Maryland, invited H.D. to present an overview of the NLog tool. He hadn't yet turned 18, but now he was on the map, and he was legitimate.

When SANS announced the presentation, "What Hackers Know About You," H.D. received top billing as the "developer of NLog, the database interface to Nmap" and an "expert on the new Nmap and NLog scanning systems that are used both by hackers/intruders and by sites to defend themselves against intrusion."

The best minds in security had been working on the new threats posed by Nmap and trying to figure out an effective way to analyze the attacks, but it took a 17-year-old hacker from Austin, Texas, to understand the exploit well enough to create a way to analyze the data. And that's what got people's attention. H.D. was using the problem to tackle the problem. Sometimes it just takes a hacker to defeat a hacker.

^ ( ) < [ ] > ( ) *^*

In March, H.D. started working as a hired hacker for CSC, and, by association, the Air Force Information Warfare Center. Fleming rushed his paperwork for a security clearance through the process and set him up to work from home via a secure virtual private network, or VPN, connection.

H.D. was still going to high school as well, and there was concern in some corners of CSC about hiring a teenager to work remotely. Most teens lack the self-discipline to work without direct supervision, they said. But Fleming was convinced that H.D. was different. He was confident that the young man he'd met that day in his office could and would do the job perfectly. And his word carried weight at CSC.

Some of the projects H.D. was asked to work on were classified above his level of security clearance, so he wasn't given any details other than the type of software he was required to produce. And if H.D. knew any additional details about the projects, he didn't let on. Regardless, he had a good idea of what he was supporting. He was being asked to write various offensive attack tools. CSC and the Air Force characterized the programs as security assessment tools. Every offensive script has a defensive application, they said. And that's true. Everything that H.D. produced was used to assess the security of military networks. Undoubtedly, the programs were also stored as part of a virtual cyberarsenal for possible use in a future information war. H.D. understood this mentality perfectly. "You have to develop exploits to test security," he said.

H.D. produced tons of code during the next nine months. But the work quickly lost its allure. Every time he got an idea and proposed something new, he said, the powers that be shot it down. "Can't do it. It's not under contract," they told him. It was symptomatic, as far as H.D. was concerned. He wasn't all that impressed with the rank-and-file hacking talent at CSC anyway. Fleming and a few of the other senior executives were sharp and knew their stuff, but most of the coders H.D. came into contact with were right out of college. Raw meat. To them, the word *exploit* meant something an unscrupulous person does to another person; it wasn't a piece of software. The squeaky-clean coders at CSC didn't have the survival instincts that H.D. had lived by for so long. And they didn't have his drive.

H.D., on the other hand, had introduced multiple security exploits for Unix and Windows systems, dubbing them open-source security tools. One such tool he called Spidermap, a collection of Perl scripts that enables hackers or security professionals to launch precisely tuned network scans. The goal of the Spidermap project was to create an integrated suite of tools for low-impact network reconnaissance. H.D. included features such as custom packet rates and scan types for each network and the ability to map multiple networks in parallel. The target users of the tool, he said, are "system administrators and network security professionals seeking a nondestructive way to inventory network services and do so in a reasonable amount of time."

H.D.'s hacking tools development was always about the technology, which is probably why he preferred to focus on speed and efficiency, as opposed to stealth. That was especially true of a program he wrote called phpDistributedPortScanner. H.D. liked to refer to it as "a joke," or a "proof-of-concept tool." Regardless of what he called it, the tool did what H.D. had designed it to do: act as a Web-based distributed port-scanning system that utilized raw speed instead of stealth. It worked, and it may have been a first.

Being first was half the fun of hacking and creating exploits. But there were few opportunities to be first at CSC. H.D. felt like he was constantly "reinventing the wheel." Then he got a chance to do something different, something interesting.

One day the San Antonio City Employees Credit Union approached CSC and asked the company to perform a security audit of the bank's networks. CSC was not in the business of performing security audits in the commercial sector, but this presented an opportunity for the firm to break new ground in a burgeoning market. Responsibility for the audit fell to Fleming, who immediately brought H.D. in on the job. CSC was getting paid $15,000 to test the credit union's systems, and the lead hacker was 17-year-old H.D. Moore.

H.D. and Fleming arrived at the credit union early in the morning. They brought laptop computers that they connected to the bank's network. Each laptop was specially outfitted with all of the state-of-the-art hacking tools, but nothing a hacker couldn't find for free on the Internet or develop on his or her own.

Dressed in a black polo shirt, black pants, and matching black shoes, H.D. fell into a trance-like state as he began the process of mapping the network and conducting preliminary reconnaissance. He didn't do anything that a hacker sitting in front of a computer outside the bank couldn't try. The bank employees went about their business, but H.D.'s concentration remained unbroken. The silence was broken only by the sound of keys tapping at a machine-gun pace and an occasional request for information from Fleming. Eventually, the entire hack fell into H.D.'s hands.

In two hours, H.D. was roaming free on the credit union's network. But the hacking continued. Clients don't like it when security pros like H.D. stand up and say they're done in two hours. So he continued to hack the network for two more days, gaining access to the deepest recesses of the financial institution's computer systems. Nothing was spared.

On the third day, H.D. turned over his results to Fleming, who put together a slide presentation for the credit union executives. He flashed on the screen the administrator passwords and the passwords belonging to some of the executives sitting in the room. He detailed for them how H.D., a self-trained professional hacker, was able to bore his way into customer accounts and just about anywhere on the network he wanted to go. If H.D. had been a professional criminal, the credit union would have been in serious trouble. The blood ran out of the executives' faces. Their confidence was destroyed, but a new resolve had been born.

When H.D. was asked how long it took him to break into the network, he answered, "a few hours." And when he was then asked why it took three days to complete the audit, he responded, "I was just trying to be thorough."

^ ( ) < [ ] > ( ) *^*

The credit union audit was the last interesting project H.D. took part in at CSC. But the experience of working for one of the biggest information technology and security consulting firms in the world instilled in him more confidence than he'd ever had before.

In December 1999, H.D. resigned from CSC to become a freelance computer and network security consultant. But the jump from hacker to business executive was not as easy as he had thought it would be. By January, he was stressed out to the max trying to pay bills and find new work.

Fleming and several other CSC executives had also left the company. But unknown to H.D., Fleming and the others had been planning a new company focusing on vulnerability analysis. But many wildcards remained to be played before the company, called Digital Defense, Inc., would become a reality, not the least of which were finding funding and developing a viable business plan.

In January 2000, Digital Defense opened for business, armed with $700,000 in venture capital and a bright future in providing vulnerability analysis and ethical hacking to credit unions. Fleming hesitated to bring H.D. on board until he was sure that the company would survive. H.D. was still a teenager, after all, and Fleming did not want his first major job change to be a leap onto a sinking ship.

But when the call came on the last Thursday of the month, H.D. was ready and willing to join. Fleming was ready to hire him, too. Business was booming.

"When do you want me to start?" H.D. asked.

"How about Monday," Fleming said.

^ ( ) < [ ] > ( ) *^*

The corporate culture at Digital Defense appears custom-made for H.D. It's laid back. With the exception of times when client meetings and on-site client visits occur, suits and ties are optional. To H.D. it's the corporate version of Gonzalo Garza: how you dress and your background take a back seat to your dedication to the technology, respect for the customer, motivation to do a quality job, and desire to learn.

H.D. often wears his black DefCon shirts to the office. Fleming gets a kick out of his "I read your email" T-shirt.

"He likes to play around and have, fun too," Fleming said recently of H.D. "Your liable to get hit in the head with a Nerf football walking into the office."

But on most days, Fleming has to tell H.D. to get some sleep. He eats, drinks, and sleeps hacking and security. But it's an enthusiastic, positive obsession, as is his dedication to loud music and an occasional beer.

H.D. Moore doesn't fit the profile of a hacker; he's an enigma who speaks in short, fast clips that somehow form complete thoughts. He sees the answers to technical problems faster than he can express them. In H.D., hacking is hard-wired. And those who know hacking and know H.D. also know that he's one of the best.

"I've been involved in the industry in many different roles over the past 20 years," said Fleming. "H.D. is among the most talented hackers that I've ever encountered, and I would place him in the top 20 worldwide."

^ ( ) < [ ] > ( ) *^*

When I asked H.D. Moore what his life might be like today if he had never discovered his love for computers, he answered that he would probably be a mechanic. The only problem, he said, is that "I'm terrible at fixing things around the house."

"The other day I tried to fix one of those programmable coffee makers, and I broke so many parts that it doesn't work any more," he said. "If it doesn't use ones and zeros and have a keyboard, I can't help you."

Afterword

Wednesday, October 21, 1998
One Month After My Conviction

Dear Diary:

My lawyer put up a good fight, but in the end it didn't matter. I was sentenced to serve 15 months, given three years probation, and buried under a mountain of fines. The judge also told me to take one last look at the laptop computer that was sitting in front of him because it would be the last real computer I would be allowed to look at for a while.

I've had a lot of time to think about what I did. And although I haven't figured out all of the answers, there's plenty of time for that. In fact, the only thing I have right now is time.

Lately, I've been thinking about what the prosecutor said as he crucified me in the courtroom. "These hacking incidents aren't pranks," he said. Apparently, some NASA systems that we used in our hacks of the Pentagon had to be shut down for 21 days, or some ridiculous amount of time, at the cost of $41,000. "This isn't like throwing spitballs at your teacher. Hackers should know that they will be caught and they will be prosecuted."

Well, he made his point, didn't he? On the other hand, I laugh when I think about all of the other hackers out there who have no fear of getting caught. What's worse is that some of them also think that even if they do slip up, nobody will take the time to actually pursue legal action against them. I thought like that once. I remember it like it was yesterday. The feelings of invincibility and the complete sense of detachment you feel when breaking and entering into a stranger's computer as opposed to their house or car. Don't let anybody try to tell you otherwise; if you want to hack a Web site and rummage through some company or government agency's computers, you might as well just break a window or pick a lock and not even bother with a computer. It's all the same. Same crime, same time.

The scene deceived me, and I blame myself for allowing that to happen. I went looking for the heroes of the hacking world, believing, like an idiot, that they would all be online just waiting to teach me and guide me. But Steve Wozniak (you know, the hacker who changed the world in a positive way when he invented the first single-board computer, the Apple) wasn't online. Instead, I met Prophet, who offered me a taste of the forbidden fruit, you might say. It's hard to get rid of that taste once you've had it. And there are thousands of others lurking about in IRC who've had a taste of it, too. It's easy for one person to realize after awhile that what they're doing is wrong. But it's not so easy when everybody around you is doing the same thing and nobody thinks it's wrong.

The court-appointed social worker didn't understand this when I tried to explain it to her. As a result, she wrote a report that said I hadn't taken "full" responsibility for what I had done, and that even after being arrested I still hadn't admitted that my actions were wrong or illegal. My lawyer had a good comeback for that. "What was he supposed to do, cry? Doesn't pleading guilty say to you that he admits what he did was wrong?" I don't think that helped my case very much.

My parents think they've raised a criminal. I can't blame them. I think about what it must be like for my dad to have to walk to the mailbox in front of the house with all of the neighbors staring. I don't think anybody on my block had ever seen a real FBI agent until the day that a dozen of them came running up to the front door of my house. The entire episode pretty much crushed my father. He came to visit me once and he looked and sounded like he was all strung out. He mumbled something about what a shame it is that any chance I might have had to get a job working with computers was now gone. No computer company would ever hire somebody who went to jail for hacking, he said. He's right. Even I know that much.

It's the realization that my career opportunities will probably be severely limited when I get out of here that makes me wonder if it all was worth it. What was The Skeleton Crew really all about? I'll tell you what it was about. It was about ego and power and being able to do things that few people could do in an environment where there was never anybody around to stop you. It started out being about the technology, but after awhile it became more about the hacker than the hack. Being in the scene today is like walking into a high school that has a gun in every locker. You never know where the next shot will come from, who will be standing behind the weapon, or if you'll be the one who gets caught in the crossfire. It's complete chaos. And it's every hacker for himself or herself. They turn on each other in a heartbeat.

Sometimes I wish I had channeled my hacking into some sort of activism. Then if I had been busted and sent here for something like defacing the home pages of companies that have bad reputations when it comes to environmental protection or human rights—two issues I'm actually interested in—then none of this would be so hard to swallow. That type of hacking actually has a purpose and can be a form of legitimate dissent.

Instead, I'm here wondering if I've added to the already tarnished reputation that hackers everywhere suffer from. I guess, in a way, I did. I became part of the problem. I lost sight of the hacker ethic because I allowed the underground to consume me. Nobody ever explained to me that the hacker ethic and the hacker underground are two different things and incompatible with each other. The observers and the experts—those who don't do, but simply watch and comment—never once said that the hacker ethic dies a slow, painful death in the underground. They failed to tell us that the hacker ethic is like a green leafy plant that needs sunlight to survive and grow. So, in many ways, I feel like I've been failed. You, the so-called experts and knowledgeable observers of the hacker culture who take pride in and intoxicate yourself with the thought of young hackers stopping at nothing to reach an objective, even if that objective involves bending or breaking laws, you don't know me. And you don't know hackers.

To my fellow hackers who are still out there: beware of the false "manifestos" that spread through the underground like a cancer. What I've just told you is how it really happens.

^ ( ) < [ ] > ( ) *^*

On the afternoon of July 27, 2001, months before I started writing this book, I arrived at FBI headquarters on Pennsylvania Avenue in Washington, D.C., for an interview with Ron Dick, the chief of the FBI's National Infrastructure Protection Center (NIPC). He agreed to talk to me about the recent outbreak of a malicious Internet worm called Code Red days before he was to hold a press conference to warn the nation about the devastating impact the worm could have on the nation's infrastructure.

When I arrived, the 23-year veteran of the FBI who'd spent five years marketing mainframe computers for Burroughs Corporation (which later became Unisys) before joining the FBI, was on the telephone. That was no surprise; he was on the telephone a lot these days. More than 250,000 computers had become infected with the Code Red worm, and it was Dick's job to stop it in its tracks and find out who was responsible.

Dressed in a dark pinstripe suit, with gold cuff links, his hair slicked straight back neatly over the top of his head, Dick waved at me to enter his office. With him sat Leslie Wiser, an investigator at the NIPC and the FBI agent responsible for nabbing Aldrich Ames, the most damaging mole in CIA history. They were brainstorming ideas about how to get the word out to the hundreds of thousands of system administrators who still hadn't patched their systems to protect against the Code Red worm.

I asked bluntly if they had any leads or if they knew who was responsible for writing and releasing the worm into the wild. Dick said he didn't. The only thing he knew for sure, he said, was that it was his job to stop it from spreading.

After we had exhausted our discussion of the Code Red worm, I asked Dick and Wiser who it was within the so-called hacker community they were really concerned about. I had just returned from Las Vegas, I said, where the annual Black Hat and DefCon hacker conferences are held, and I was less than impressed with the maturity and professionalism of many of the attendees. I had expected something very different, I said. Where were the dark, brooding, leather-clad killers who we've been reading about in the press all these years? I asked. Sure, there was a significant amount of misguided, ill-informed, anti-establishment opinion to be found there, but for the most part these so-called black-hat hackers were just teenagers, kids.

Dick assured me that these were not, in fact, the hackers who kept officials in Washington up at night. Script kiddies, as they were known, were not the focus of the NIPC, he said. The NIPC took action against this group when necessary, but they were not the main threat to national security. Foreign intelligence services that employ battalions of hackers, terrorist groups or sympathizers that seek to bring down major portions of the nation's Internet-based infrastructure, and organized crime rings were the groups that official Washington cared most about. These groups employ professional hackers who are clever enough not to get caught red-handed, and who are not interested in making a big scene. They want to enter systems unseen, map out networks, collect intelligence information, install backdoors, and even create "zombie" machines that can be ordered remotely at any time to take part in massive denial-of-service attacks. These are the types of hackers we really worry about, Dick said.

Then, six weeks later, something happened that both validated and changed everything Dick had told me.

On the morning of September 11, I was walking out of a Starbucks coffee shop and had barely reached my car when a woman, wide-eyed and shaking with fear, ran up to me and told me to turn on my radio.

"They hit the World Trade Center!" she cried.

"Who hit the World Trade Center?"

"They think it was a terrorist attack," she said, her voice cracking under the strain of the moment. "They flew two planes into the towers!"

As an American, the news was an assault on my soul. It was a horrible and mind-numbing experience that would become worse as the day wore on. But as a reporter, word of the attack meant something far different. It meant that the slate had been wiped clean, new priorities for the reporting week had been set, new challenges lay ahead, and a host of new, disturbing questions awaited answers. The day before, I had planned on writing about a new Pentagon report detailing attempts by foreign intelligence agents to obtain information about U.S. high-tech systems. I had also planned on writing a story about a new technology whose creators claimed could make a system's hard drive "invisible" to would-be hackers and crackers. But with a quick call on a cell phone to my editors at *Computerworld*, that all changed.

During the three months that followed the attacks on September 11, I reported and wrote more than two dozen stories on the increasing threat of cyberterrorism. Like the attacks against the World Trade Center and the Pentagon, cyberterrorism wasn't a new threat that people were hearing about for the first time. But since the attacks, it had taken on new meaning, new significance. The nation relied on computer networks for everything, from banking and finance to electric power and 911 emergency systems. Terrorists were, in fact, also interested in hitting America with digital bombs capable of ripping through the infrastructure and causing widespread financial damage, disruption, and mass confusion. Osama bin Laden himself had reportedly made a statement to an Arab newspaper claiming this to be one of the al Qaeda terrorist organization's goals.

But something else happened during those three months. The fear of terrorism, especially the unknown forms that terrorist attacks might take in the future, led officials in Washington to redefine the meaning of the word *terrorism*. Suddenly, you didn't have to be John Walker fighting alongside the Taliban and Osama bin Laden in Afghanistan to be considered a terrorist. According to the U.S.A. Patriot Act, signed into law on October 26, hackers can be terrorists, too. And your age doesn't matter. But what you do does matter.

All of these events occurred as I was conducting interviews and research for this book. As it turns out, I couldn't be more thankful for the opportunity to write *The Hacker Diaries* during this particular time in our nation's history. It is during such times of sweeping, lightning-fast changes in perception that we need more than ever to find the truth in individual experience. During the course of the past six years, I, too, had come to view the vast majority of the hacker community as a bunch of common criminals, with nothing better to do with their time than break into computers and deface Web sites. But it took writing this book and talking to real hackers to see how wrong I had been about the majority of hackers. All hackers are not the common criminals we read about, and they are certainly not terrorists.

I have come to admire the vast majority of the individuals interviewed for this book. Their deep devotion to the advancement of technology and their desire to push technology's limits are truly unique. Another admirable characteristic of some of the hackers profiled in this book is the desire to use their hacking skills in support of a social cause. Teenagers have always been the vanguard of healthy political protest in America. And rules often get broken during such protests, especially when the mainstream ignores real problems. Hacking and defacing Web sites with messages that take a stand on important social and political issues is, therefore, not always a bad thing.

However, computer hacking today represents a significant departure from the traditional methods employed by American teenagers to provide an outlet for their curiosity, angst, and dissatisfaction with the status quo. Whereas young people used to conduct school sit-ins to protest government policies or perceived injustices, pull fire alarms as a practical joke that would get them out of class, or shoot spitballs at a new teacher or classmate, today they use the Internet to single-handedly challenge the authority of powerful corporations and government agencies. But the damage that teenage hackers can cause and the seriousness of their actions are far greater than the repercussions of a practical joke.

More significant, however, is the character that many of the hackers in this book have shown. While many started and learned their trade by bending and sometimes breaking the rules, few, if any, of the hackers I have come to know while writing this book did so out of a pure desire to inflict damage or personal injury. And all, with the exception of Mafiaboy and Pr0metheus, have come to recognize the error of their youthful ways. They're human beings who have made mistakes. And all of them certainly didn't grown up with silver spoons in their mouths, nestled away in quiet suburbia with caring, loving, stable families. The fact that many of them have since focused their talents on lawful activities, even, as in the case of Genocide, of ridding the Internet of child pornographers, is truly commendable.

Teenage hacking, particularly the act of defacing public and corporate Web sites, is a cultural phenomenon that knows no borders. The roots of teenage hacking run deeper than any one celebrity hacker or group. As a result, it is a phenomenon of the information-age culture, not of any one country or geographical area. Thousands of Web sites run by governments, businesses, churches, schools, and nonprofit organizations are defaced every year. Teenagers of all ages, from all walks of life and from every part of the country, are getting involved in hacking.

There is, however, another side to the teenage hacker underground. The Internet has provided an outlet for an alarming level of anger and disconnection among teenagers. That anger and frustration has been demonstrated in a frightening wave of school violence in recent years, and increasingly the anger and sense of disconnection felt by teenagers is manifesting itself in hacking and Web site defacements that are violent, racist, vulgar, and fervent in their anti-establishment, anti-government leanings. Likewise, the resentment felt by today's teenage hackers has also led to an increase in the destructiveness of teenage hacking exploits. Whereas teenage hackers used to be interested in exploring and in sharing information, today a growing number of them appear to be interested in destroying and blocking information flow—a trend that is at odds with the essence of the hacker culture.

It is also clear from my conversations with hackers and research into some of the high-profile cases involving teenage hackers that parents have no idea what their kids are getting into online. They concern themselves with pornography and fail to seek education in how to detect whether or not their sons or daughters are involved in hacking. After all, many teenage hackers today don't have a history of getting in trouble with the law. Parents are usually comfortable leaving them alone in their bedrooms sitting in front of their computers. They're in their rooms working at their computers, not out getting drunk or getting high. In most cases, no alarm bells are sounding.

The public education system also shares responsibility for whatever mischief is being caused by teenage hackers. The public education system does not teach kids responsibility and responsible use of computer resources. The absence of such ethics courses in the public school curriculum represents a significant failure of the education system. Most of the teenagers interviewed for this book have little or no respect for the level of technical competency demonstrated by their computer science teachers. Many teenagers become bored with substandard teachers who know less about computers than they do. As a result, they seek out special assignments to feel challenged. And sometimes when those special assignments don't materialize, they turn to hacking on their own. Public schools in the U.S. have become the weak link in the effort to prevent teenagers from getting involved in illegal hacking activities. And so have parents.

If you think parental apathy is not a factor in teenage hacking, think again. Consider how easy it was for two upper-middle-class teenagers in Colorado to stockpile weapons in their bedrooms right under the noses of their parents. If parents cannot find guns and explosives in the rooms of their kids, how hard do you think it is for parents to uncover the online activities of their sons or daughters? Despite the spread of technology throughout our society, many parents still have a hard time figuring out how to turn on a computer, never mind knowing how to spot signs that their teenager has been visiting hacker Web sites or exchanging communications with other, more dangerous, members of the underground.

For example, on December 1, 2000, Raymond Torricelli (a.k.a. Rolex and a member of the hacking group #conflict) pleaded guilty to hacking into computers belonging to NASA's Jet Propulsion Laboratory in Pasadena, California, and San Jose State University in 1998 when he was 18 years old. He also admitted to taking credit card information from the computers to set up illegal telephone services and downloading password files to gain free access to Internet accounts. However, the majority of the evidence used against Torricelli was obtained through a search of his personal computer. In addition to thousands of stolen passwords and numerous credit card numbers, investigators found transcripts of chat-room discussions in which the teenager discussed breaking into other computers, obtaining credit card numbers belonging to other people, and using those numbers to make unauthorized purchases. Officials also uncovered evidence of a plan to electronically alter the results of the annual MTV Movie Awards.

Initially, I thought that if parents just took the time to learn what their teen-ager's online screen name was they would be a step ahead of the game. However, during the course of interviewing the hackers who appear in this book, I soon realized that most hacker nicknames are chosen either for shock value or as an inside joke. There seem to be few cases where a screen name actually indicates the type of person a hacker really is. Most do not live up to the shocking hacker handles they choose.

Steve Wozniak, the famed inventor of the first Apple computer, was a hacker. During the early 1970s, he learned the finer points of "phone phreaking" from another famous hacker named John Draper (a.k.a. Captain Crunch). Mesmerized by Draper's ability to use a plastic toy whistle to replicate the tone that was used by the public telephone system to authorize long-distance calls, Wozniak began building little blue boxes to produce the tone and started making his own calls. One of his first phone phreaking experiences involved making a free call to the Pope at the Vatican in Rome. But Wozniak's dedication to pushing the boundaries of computing technology would soon lead to the first Apple desktop computer, which helped change the world. He elevated hacking to a legitimate scientific pursuit and took great pride in solving problems with fewer lines of code or less hardware than the next guy. Through his work and the work of the founding fathers of the hacker movement, hacking produced benefits for society and revolutionized the business world. It's hard to imagine where the computer industry would be today had it not been for hackers like Wozniak.

Today most hackers consider hacking a public service to the Internet community. The argument goes something like this: Hackers expose security vulnerabilities in widely used commercial software, and therefore, their exploits help make the Internet a more secure environment for consumers. These discoveries are fueled by the hacker's curiosity, desire to push technology to its limits, and disdain for large software development companies that sell applications with woefully inadequate security protections. In addition, when hackers discover new vulnerabilities, they force online companies to take swift action to plug the security holes in their software that might otherwise go unfixed. If it weren't for hackers, companies would never take the necessary steps to protect customer data, and security technology would founder. That remains a core belief of many in the hacker community.

These same hackers also believe in the depths of their souls that hacking is justified because the Internet and the information on the Internet should be a free resource for all. Information and knowledge cannot become the exclusive property of big businesses and governments, they argue. Therefore, hacking into computers and the downloading and copying of information is not a crime in the minds of these information-age intellectuals. Hackers look at themselves as Internet-age Robin Hoods, stealing from the information rich to give to the information- and connectivity-starved poor. Their aim is to open up and expose information held closely by corporate America and the government and expose the truth. The world's knowledge belongs to the world, not a select few with the money and political influence to claim ownership of it. The freedom of information and knowledge is another core belief of the hacker community.

It was this belief that all information and knowledge should be free to all who want it that landed legendary teenage hacker Kevin Mitnick in federal prison in January 1995. In addition to a series of pranks and an episode in 1981 that involved physically breaking into the offices of a Pacific Bell switching station and stealing technical manuals that outlined how the system worked, Mitnick's Internet high jinks gradually progressed to include more serious crimes. In 1988, for example, Mitnick, who was 25 at the time, secretly monitored the e-mail communications of MCI and Digital Equipment Corporation. He then stole, or in his words, simply "copied," $1 million worth of software. He received a one-year jail sentence for these crimes. But it didn't end there.

Mitnick continued hacking and gradually increased the boldness and severity of his crimes. In 1993, for example, the California State Police issued a warrant for his arrest and accused him of wiretapping telephone calls from the FBI to the California Department of Motor Vehicles. He then used law enforcement access codes to hack his way into the DMV drivers' license database. When all was said and done, the companies damaged by Mitnick's hacking activities claimed to have lost $291.8 million. Mitnick would spend almost five years behind bars in federal prison for failing to realize that the world around him had changed. Information had become as valuable as cash, and companies had started to treat it as such. Information and intellectual property wasn't free for the taking as the hackers argued that it should be. Information was the lifeblood of the economy, upon which depended the standard of living and way of life of millions of people. Likewise, personal privacy was fast becoming a hot-button issue in an age when it was possible for people like Mitnick to violate the privacy of thousands of innocent, law-abiding citizens.

Mitnick, however, never changed his opinion of what he did, and his arrest gave rise to an entire generation of hackers who shared (and still share) his myopia.

Many former hackers also agree that while some teenage hackers represent the next generation of computer security movers and shakers, their motivations are strikingly different than those of their predecessors. Whereas the early hacker movement was all about opening up access to information and defending the Internet as a resource of the people that is free from big government and big corporations, teenage hackers today are all about blocking access to information and the Internet. Conducting a distributed denial-of-service (DDoS) attack, for example, hurts everyone, not just the companies and governments that use the Internet to conduct business. Likewise, destructive viruses and worms are simple to create and do not discriminate. They have very little value for hactivists or hackers who want to demonstrate their technical prowess. Yet teens and amateurs (and, admittedly, foreign hackers with murky backgrounds) continue to toy with them, to the tune of about $15 billion in damage to the U.S. economy every year.

Although the hacker scene appears to be changing for the better, as can be seen by the appearance of hackers such as Anna Moore, Willie Gonzalez, and others, those who hack out of a desire to cause chaos and disrupt the flow of information for all Internet users far outnumber those who are interested in hacking as a legitimate scientific pursuit.

On the other hand, some security experts who talked to me on condition of anonymity said they fear and deplore those who claim to be legitimate security consultants while at the same time taking part in the creation of damaging viruses and worms and then letting them loose in "the wild." These former hackers understand that some unwitting teenager will almost certainly try to play with these malignant creations and send them out across the Internet. After that, it's money in the bank for the security firm.

In other cases, hackers and computer security professionals with close ties to the hacker community expressed concern about becoming the target of reprisals from the hacker community for speaking the truth. This was especially true for a few individuals who said they are aware of computer security professionals employed by major security firms who continue to actively hack and discuss their latest exploits and vulnerability discoveries with other hackers on various IRC channels. These hackers have gone from spending 10, 12, or 14 hours per day hacking to, in the words of one security administrator with ties to the underground, "creating their own buzz" to benefit their new careers in security. Teenagers, therefore, are not the only members of the under-ground causing real problems for Internet users.

Hackers, whether teenagers or adults, are real people with real problems and face the same pressures that most ordinary people do. They are not the all-knowing evildoers depicted in the media. Despite their personal challenges, however, the Internet truly is a more secure place because of the genius of hackers.

In the end, however, we are left with the vexing questions that started us off on this journey: *who are teenage hackers, what motivates them to hack and sometimes bend or break the law, and what do they think of themselves and the underground society of which they are a part?*

Teenage hackers are all of us; everything we are, everything we are not, everything we never intended for ourselves, and everything we wish we could be. They are our sons and daughters; our neighbors; our friends and co-workers; and the kid shoveling snow for the elderly, raking leaves on the weekends, or playing sports.

Teenage hackers are the great explorers of the Information Age. Some will stop at nothing to discover the possible in that which others say is impossible. These are the minds that have given the world great things, and the minds, unblemished by wisdom, that are still courageous enough to see the world in terms of right and wrong. And these are the minds that have the unique ability to think digitally, the minds that breathe life into silicon, though yet still inexperienced in the ways of the world and in need of a moral compass.

The teenage hacker underground is home to a great conspiracy that is being planned and carried out from small, sloppy bedrooms, high school computer labs, and mall food courts all over the world. The goal of the conspiracy is to change the world for the better and, if necessary, to shed light on what doesn't work so that it can be torn down and rebuilt correctly. The goal is to change what needs to be changed: their lives, their world, or the Internet. And in a world where nothing is beyond hacking, they just might do it.

A

Two Decades of
Teenage Hacking

Hacking traces its roots back to the legitimate scientific pursuits of technologists in the 1950s and 1960s. However, with the rise in the 1980s of celebrity outlaw hackers like Kevin Mitnick and the spread of technology, hacking has become a national past time for many curious and rebellious teenagers. It has also worked its way into the mass media, becoming the focus of movies and a centerpiece of our popular culture. Here's a look at some of the major hacking events and personalities of the last 20 years.

1981

—Kevin Mitnick, 17, is convicted of stealing computer manuals from an office at Pacific Bell. Mitnick would go on to become America's most wanted hacker.

1983

—The movie *War Games* is released. It stars Matthew Broderick as a teenage computer hacker who accidentally hacks his way into the Pentagon's nuclear command and control system and almost starts World War III.

—Seven Milwaukee-based teenage hackers, known as the 414s, are busted in one of the first major hacker stings to gain national notoriety. The 414s, named after the local area code where they lived, were charged with and convicted of breaking into more than 60 systems, including Security Pacific National Bank and the Los Alamos National Laboratory.

1987

—A 17-year-old-high school dropout named Herbert Zinn, known in the hacking world as Shadow Hawk, admits that he broke into AT&T networks. The teenager worked from his bedroom in suburban Chicago.

1988

—Twelve-year-old **Willie Gonzalez** witnesses his first hack while answering telephones with the teenage son of his father's boss. It marks the beginning of his quest to become a hacker.

1989

—A 10-year-old named **Joe Magee** from Philadelphia and a 12-year-old known as **Noid** from a suburb of Chicago discover their parents' VCRs and simultaneously embark upon journeys that will eventually introduce them to computers and take them through the back alleys of the Internet. There, they will enter a mysterious, uncharted universe known as the hacker underground.

1990

—A nationwide hacker crackdown nabs teenage members of the notorious hacking groups known as the Masters of Deception (MOD) and the Legion of Doom (LOD). The teen hackers are responsible for the famous Martin Luther King, Jr., Day crash of the AT&T long-distance telephone network. The hackers would be indicted in 1992.

—**Genocide**, a 14-year-old from Fairbanks, Alaska, discovers the world of hacking after he steals time on the local university network using his mother's account.

1992

—Eleven year-old **H.D. Moore**, who will go on to become one of the best hackers in the world, learns how to code in BASIC and begins creating his own games.

1994

—British authorities arrest a 16-year-old boy after a U.S. Air Force investigation proves that he was involved in breaking into the systems at NASA and Griffith Air Force Base.

—Seventeen-year-old **Willie Gonzalez** routinely logs 16 to 18 hours a day in front of the computer looking for more challenging systems to hack.

1995

—The movie *Hackers* is released. It paints a stereotypical picture of the outlaw teenage hacker and leads to a series of Web page defacements by angry hackers, including the movie's home page, created by Metro-Goldwyn-Mayer Studios, which released the film.

—Kevin Mitnick is arrested shortly after breaking into a system at the San Diego Supercomputer Center.

—The Genocide2600 group is formed, with **Genocide** as its leader.

—Nine-year-old **Anna Moore** discovers her first chat room on the Internet.

1996

—Christopher Schanot, 19, is arrested and released under house arrest after breaking into military networks, a defense contractor system, Sprint, a credit-reporting agency, and a telephone company.

—The **Genocide2600** group's rampage through university networks hits high gear. The group's Web page becomes a major source of hacking texts and software tools.

1997

—**Genocide** writes his hacker's manifesto titled *The Social Base of the Hacker*.

—Eleven year-old **Star Road** (a.k.a. **Anna Moore**) becomes a combatant in the online hacker wars known as the Nuke Wars.

—**H.D. Moore** begins actively "war dialing" for modem connections and discovers serious vulnerabilities in the application code used by America

Online, Yahoo!, and Microsoft. The companies are thankful that he reports the problems and offer to provide the young hacker with recommendations when he applies for jobs.

—A 14-year-old kid nicknamed **Explotion** begins to see the "right and wrong in his parents" and delves deeper and deeper into computers.

1998

—An 18-year-old Israeli hacker named Ehud Tenebaum and two California-based teenage hackers are arrested for conducting what Defense Department officials would describe to Congress as the "most organized and systematic attack" against Pentagon networks to date.

—A teenager whose name is withheld by authorities is charged with hacking into an FAA control tower at Massachusetts's Worcester Airport and disrupting vital systems for six hours.

—Members of the L0pht hacker group testify before a Senate committee, warning of serious security weaknesses on the Internet. They claim it is possible to shut down the Internet in half an hour.

—An 11-year-old hacker and future member of World of Hell nicknamed **Kron** begins his hacking experimentation. He squeezes his hacking in between soccer practice and hanging out with friends.

—The Navy "Shadow" group discovers that a high level of stealth scans are being conducted against Pentagon networks from hackers all over the world.

—**H.D. Moore**, 16, writes a program called NLog that enables the "Shadow" team to conduct analysis of the new stealth scans. He's later invited to present his findings to a live meeting of the System Administration, Networking and Security (SANS) Institute.

1999

—Eric Burns, a 19-year-old hacker known as Zyklon, is arrested for hacking into the White House Web server. Burns is sentenced to 15 months imprisonment and three years of supervised release and ordered to pay $36,240 in restitution.

—**Mafiaboy** hits one of his first victims, Sisters High School in the small town of Sisters, Oregon, which will later play a significant role leading to his capture.

—Kevin Mitnick is sentenced to three years and ten months in prison and credited with time served.

—**Anna Moore** masters the Linux operating system and begins to code in the C programming language.

—**FonE_TonE**, a 15-year-old hacker and future member of the World of Hell gang, publishes two informative hacking texts on the importance of deleting log files, Cisco router security, and how to defeat encryption. He finds some of his information on the **Genocide2600** Web site.

—Seventeen-year-old **H.D. Moore** is recruited by Computer Sciences Corporation to work on classified Air Force information-warfare programs.

2000
—On January 21, 2000, Kevin Mitnick is released from federal prison.

—Dennis Moran, 18, (a.k.a. Coolio) is arrested for redirecting legitimate Web traffic away from the Web site of RSA Security and twice breaking into the anti-drug site Dare.com. Moran is sentenced to nine months in a New Hampshire prison and is ordered to pay his victims $15,000.

—Chad Davis, 19, of Green Bay, Wisconsin, a member of the hacker group Global Hell, is sentenced to six months' imprisonment and three years of supervised probation and ordered to pay $8,054 in fines for hacking into and damaging the home page of the U.S. Army.

—Patrick W. Gregory, also known as MostHateD, receives a two-year prison sentence and three years supervised release time and is ordered to pay $154,529.86 in fines. He admits to being a founding member of Global Hell at the age of 17 and is found to have stolen credit card numbers and personal identification numbers from innocent people and used them to tap into teleconferencing services of AT&T, MCI, and Sprint to the tune of more than $22,000 in unauthorized charges.

—An unidentified 15-year-old and 20-year-old Raymond Torricelli are arrested for breaking into NASA computers. The teen acknowledges that another hacker was tutoring him online. Police find 76,000 passwords on Torricelli's computer.

—Russell Sanford, 18, a member of the hacker group HV2K, is sentenced to two years in prison and ordered to pay $45,856.46 in fines stemming from hacking and defacing Web sites belonging to the U.S. Postal Service, the State of Texas, and other private companies.

—A 15-year-old Canadian teenager who calls himself **Mafiaboy** is arrested in Montreal and charged with 58 counts of illegal access, mischief to data, and other violations stemming from the February denial-of-service attacks that took down the Web sites of Yahoo!, CNN, E-Trade, Amazon.com, E-Bay, Dell Computer, and others.

—Two high school students in Michigan, William G. "Greg" Lulham, 17, and Lloyd S. Dilley, 18, are charged with two felonies each stemming from their alleged attempt to hack into the Corunna School District's computer system. The teenagers are cousins and were considered "model students."

—**Explotion** leaves public school and enters a special technology school, where a teacher endorses his hacking experimentation.

2001
—**Mafiaboy**, whose real name has been withheld, is sentenced to eight months in a juvenile detention center.

—In March, **Cowhead2000** forms the World of Hell hacking group. The gang that starts as a joke quickly turns into one of the most prolific Web site defacement groups of all time.

—In June, **RaFa** joins World of Hell and introduces the group to political and social defacement messages.

—In November, federal authorities raid the home of **Cowhead2000**, and his hacking career comes to an abrupt halt. Other members of World of Hell try to revive the group, but become frustrated with what they see as a hacking scene in decline.

—**Pr0metheus** leaves the hacking group known as PoizonB0x to form Hacking For Satan and begins a campaign of defacing Christian and other religious Web sites to promote Satanism.

—Fifteen-year-old **Anna Moore** (a.k.a. **Starla Pureheart**) wins the CyberEthical Surfivor contest at the annual DefCon hacker conference in Las Vegas. It is a sign of the times.

—Nineteen-year-old **Explotion** acknowledges writing a script that allows him and his friends to rig an online survey sponsored by one of his favorite bands.

B

Making Headlines Over the Years

1994
Hackers stay a step ahead, *Computerworld*

1995
Hollywood puts hackers on pedestal, *Computerworld*
FBI nabs notorious hacker, *Computerworld*
Six hackers stung by undercover cyberspace, *Computerworld*
Underground tools aid fledgling hackers, *Computerworld*
Gotcha! A hard-core hacker is nabbed, *U.S. News & World Report*

1996
Computer hacker plants porno on Air Force Web page, *CNN*
'Tis the season for hackers, *Computerworld*
Holidays bring out hackers, *Computerworld*
Hackers deface Air Force Web site, *Computerworld*
Hacker defaces Justice Department's Web site, *Computerworld*
Hackers step up attacks, *Computerworld*
Few gains made against hackers, *Computerworld*
University hacker to be hunted on the Internet, *Daily Telegraph (U.K.)*
Hackers go into killer mode, *Daily Telegraph*

1997

Teenager admits $100,000 credit card rip-off, *Associated Press*

Men in black (hats) identify future hacker targets, *Computerworld*

Phone hackers dial-up trouble, *Computerworld*

Hackers jump to their own defense, *Computerworld*

Hackers target Microsoft's Web site—again, *Computerworld*

Hacker pens guide to Internet security, *Computerworld*

Military beefing up its hacker defenses, *Computerworld*

Hacker pleads guilty to defrauding AOL, *Computerworld*

AOL hacker gets probation, home confinement, *Computerworld*

Congressional panel hears troubling news on hackers, *Computerworld*

Hackers' dark side gets even darker, *TechWeb News*

Ontario boy, 14, charged as hacker for breaking into more than 500 sites in less than a year, *Vancouver Sun*

1998

Thousands of passwords accessed by cyber prowler, *Associated Press*

CIA head foresees better hackers, *Associated Press*

Teen hacker faces federal charges, *CNN*

Hacker suspect called "damn good…and dangerous," *CNN*

Hackers disrupt N.Y. Times site, *Computerworld*

IT security opportunities: The hackers among us, *Computerworld*

Want to prevent break-ins? Just ask a hacker, *Computerworld*

Hacks gain in malice, frequency, *Computerworld*

Feds struggle in race with hackers, *Computerworld*

Hackers team up for large-scale attacks, *Computerworld*

Hackers spoof Computerworld, *Computerworld*

Teen hackers may be tried as adults, *United Press International*

Terrorism at the touch of a keyboard, *U.S. News & World Report*

Analyzer: Primed to kill since 6, *Wired*

Teen crackers admit guilt, *Wired*

FBI mounts big crackdown on small-town teens, *ZDNet News*

1999

FBI on offensive in "cyber war," raiding hackers' homes, *CNN*

Interior Department hit by hackers, *Computerworld*

Lovesick hacker hits Microsoft site, *MSNBC*

FBI tries to crack Global Hell ringleader, *Newsbytes*

Hackers vow retaliation against FBI pressure, *Newsbytes*

Israeli boy, 14, hacks Saddam off the Internet, *Sunday Telegraph*

Web of disruption—Hacker, 16, suspected of wide campaign, *The Montreal Gazette*

Suddenly hackers are sexy, hip and evil, *The New York Times*

Five arrested for hacking into high school system, *Flagler Palm Coast News Tribune*

2000

MafiaBoy arrested for distributed denial-of-service attacks, *Associated Press*

Hackers hardly huggable, *Chicago Tribune*

NOW HIRING: HACKERS (TATTOOS WELCOME), *Chicago Tribune*

Feds warn hackers, then ask them for help, *Computerworld*

Hospital confirms copying of patient files by hacker, *Computerworld*

Republican Web site hit by hacker, taken off-line, *Computerworld*

Hacker/Fed tensions abound at DefCon, *Computerworld*

Experts: Hackers may be infecting thousands of Windows PCs, *Computerworld*

Feds ID hacker who allegedly stole more than 485K credit card numbers, *Computerworld*

Justice Department says new laws needed to track hackers, *Computerworld*

Teen hacker arrest masks true Net peril, *E-Commerce Times*

Teen first to serve hacking time, *InfoWorld*

Mitnick barred from computer columnist job, *Newsbytes*

AtStake jilts Phiber Optik: The corporation formerly known as the L0pht courts Mark Abene, balks at his hacker past, *SecurityFocus*

Probe of hacker nets second suspect: His father, *The Washington Post*

Teen hacker to serve time, *Wired*

2001

At Black Hat, ties seen tightening between hackers,
legal officials, *Computerworld*

Black Hat highlights real danger of script-kiddies, *Computerworld*

Angry hacker releases customer data of Wyoming ISP, *Computerworld*

Federal systems increasingly falling prey to hackers, *Computerworld*

Hacker hires don't interest most businesses, *InformationWeek*

Hackers, not terrorists, major concern, *InternetWeek*

Businesses battle growing plague of hackers, *Los Angeles Times*

Computer hacking a form of terrorism, *Los Angeles Times*

Satanic hacking group targets church sites, *Newsbytes*

Unholy hackers deface Vatican Web site, *NewsFactor Network*

Hacking of Web game EverQuest linked to local teen, *The Seattle Times*

Fighting evil hackers with bucks, *Wired*

As ethicists, they don't hack it, *Wired*

Senator targets school hackers, *Wired*

Call them kiddies? Watch out, *Wired*

C

Hacking on the Web

Individual Hackers

H.D. Moore
www.digitaloffense.net

The personal Web site of H.D. Moore, where you'll find links to all of his open-source exploits, worm code, analysis of security issues, and all sorts of other goodies.

John Vranesevich
www.antionline.com/jp

The Web site of the founder of the hacking Web site AntiOnline.com, thought to be one of the best hackers in the world.

Kevin Mitnick
www.kevinmitnick.com

The official Web site of legendary hacker Kevin Mitnick, including information about his hacking exploits, the government's case against him, his time in federal prison, and the government's recent push to revoke his radio license.

Lord Blackbane's Underground Lair
http://skyscraper.fortunecity.com/chaos/398/

Offers a collection of "war scripts" and hacking files, as well as a section on "anarchy, terrorism and lock picking."

Steve Wozniak

www.woz.org

The official Web site of esteemed Apple computer inventor Steve Wozniak. Includes links to information about his "blue boxing" days and the story about how he met the legendary phone phreaker "Captain Crunch."

Hacker Groups

Arab Hackers

www.arabhackers.org

A Web site that sells itself as "an effort to provide a homestead for Arab computer professionals and hobbyists across the world" and an online community that's "not about illegal breaking into systems, or trading warez or crackz." However, you'll find a message board with requests for hacking assistance from pro-Arab hackers and plenty of anti-American rants.

Chaos Computer Club

www.ccc.de

Home page of the world-famous hacker club based in Germany. A mix of English- and German-language information, most of which is very general.

Cult of the Dead Cow (now @Stake)

www.l0pht.com

The Cult of the Dead Cow (cDc) is best known as the group that authored and distributed Back Orifice, an open-source software product that allows a hacker to take over a remote computer. However, the group has since gone legitimate under the auspices of @Stake, a security consulting firm. That's where you'll end up with this link.

Genocide2600

www.genocide2600.com

The home page of the Genocide2600 hacker group. Links to individual member home pages, hacking texts, and other goodies.

Hacking Information

2600: The Hacker Quarterly
www.2600.com
   The home page of the hacking and phreaking magazine *2600*. Includes various links to information of interest to all hackers.

AntiOnline
www.antionline.com
   Security and hacking news, online discussions, and links to various exploit downloads and text files for learning.

Attrition.org Defacement Archive
www.attrition.org/mirror/attrition
   Although it's no longer active, the Attrition Web site archives hundreds of hacker defacements going back to 1995.

Checksum
www.checksum.org
   A self-described site for hackers by hackers. An interesting collection of hacking tools and downloads.

DefCon
www.defcon.org
   The home page of the infamous annual hacker conference held each year in Las Vegas. Includes links to conference program material and hacker information and organizations.

H3C
www.hack3r.com
   Calls itself the source for security information and the underground. Links to hacker war games, bulletin boards, and a collection of "informational papers" on security and hacking.

Hackers.com: Next Evolution Hackers
www.hackers.com
Offers a good section of links on how to start hacking, including texts and exploits.

Hack in the Box
www.hackinthebox.org
News, documents, security tools, and free discussion of hacking and security topics.

Sonic Hacking Community
www.sonichacking.org
Lots of hacking information, downloads, utilities, and a good help section for newbies.

The Hacker's Defense Foundation
www.hackerz.org
A Web site that will teach you how not to get busted and what to do if you do get busted. A Not-for-Profit foundation "committed to the advancement of the hacking community, through education, of the social, political, and legal implications of the uses of technology, and seeks to enlighten the public and law enforcement about the hacking community, through education, that hackers are not the lawless goons that law enforcement, the news media, and Hollywood would try to portray them as." However, "the Hacker's Defense Foundation does not condone, support, or defend blatant criminal acts."

The Hackology Network
www.hackology.com
A Web site run by self-described "controversial programmers," not hackers. A collection of offensive hacking tools that obviously can be used for good things (security) or bad things (you figure it out).

The Library
www.ladysharrow.ndirect.co.uk/library/
   A listing of complete books and files all available in HTML format. You'll find everything from the history of the Internet to hardware reference books.

The Texts Library
www.securitywriters.org/texts.php
   Text files on hardware, hacking, coding, networks, Unix, phreaking, and crypto.

Security Information

Computer Emergency Response Team
www.cert.org
   A clearinghouse for computer security vulnerability analysis and incident analysis at Carnegie Mellon University.

FBI National Infrastructure Protection Center
www.nipc.gov
   The home page for the cyberthreat and warning arm of the FBI. Includes links to FBI threat warnings and analysis of security vulnerabilities.

HackerWatch
www.hackerwatch.org
   Home of the anti-hacker community. A repository of firewall logs and other attack information from which members can analyze trends and improve defenses.

Internet Relay Chat (IRC) Help
www.irchelp.org
   Confused about how IRC works? All the information you'll ever need about how to use IRC and where to find like-minded people to chat with are here.

Phrack Magazine
www.phrack.org
The hacker magazine written for and by the hacker community.

Report a Cybercrime
www.cybercrime.gov
The home page of the Department of Justice's Computer Crimes and Intellectual Property Division. Includes links to case summaries and a way to report cybercrimes to the feds.

Teen Programmers Unite
www.tpu.org
An organization formed in 1996 for teenage computer programmers worldwide. Its goal is to help people who fit into this growing category meet each other, exchange ideas, perhaps work together on something, and learn. Very technical.

W00w00 Security Development
www.w00w00.org
A nonprofit security research group with members in 11 countries. Offers a good set of links to texts and files archives, including exploits.

INDEX

References to figures and illustrations are in italics.

Numbers and Symbols
#conflict hacker group, 189
@Stake, Web site, 208
0-day exploits, 54
2600 hacker group, 36, 135–136
2600: The Hacker Quarterly, 11–12
    Web site, 209
    See also Genocide2600
414s, 196

A
Aaron. *See* Noid
Abene, Mark, 35
Acid Phreak. *See* Ladopoulos, Eli
activism, 184, 187
    hactivism, 96
al Qaeda organization, 112, 187
Alexu, 10, 12
America Online, vulnerabilities,
    173–174
    AOHell, 21
Ames, Aldrich, 185
AntiOnline, Web site, 209
AOHell, 21
AOL. *See* America Online
Apocalypse, 43–46, 49–50
Arab Hackers, Web site, 208
Aracnet, 22
Astroboy, 10, 12
Attrition.org Defacement Archive,
    Web site, 209

B
Back Orifice, 176
BBSs, 34–35, 42–43, 147
    Revenge, 43–46

bewm, 149, 150
bin Laden, Osama, 112, 187
Black Hat, 185
blue boxes, 5, 190
blue screen of death, 149
bots, 131
Broderick, Matthew, 37
Burns, Eric, 198

C
Calce, John, 58, 85–86
Captain Crunch. *See* Draper, John
cDc. *See* Cult of the Dead Cow
Cerf, Vint, 78
Chaos Computer Club,
    Web site, 208
Checksum, Web site, 209
child pornography, 26, 188
Christian Web sites, hacks against,
    90–101
chronology, 195–201
    headlines, 203–206
Clinton, Bill, 77–79, 113
CNN attack, 70
Code Red, 185
Columbine High School, 1–2
Computer Emergency Response
    Team, Web site, 211
Computer Intrusion Squad, 60
Computer Sciences Corporation,
    166–170, 178–180
Coolio. *See* Moran, Dennis
Cowhead2000, 109
    chronology, 200
    computer setup, 111–112

defacement of Sewon
Teletech, 114
FonE_TonE, 121
founding of World of Hell,
114–116
raid by federal agents,
110–114
See also World of Hell
Crack, 8–9
Cult of the Dead Cow, 83
Web site, 208
Currie, Robert, 82, 83–84, 89
CyberEthical Surfivor contest,
127–129, 140, 201
See also DefCon
cyberterrorism, 186–187
CYNOSURE, 50–51

D

DA Chronic, 21
Datek attack, 70
Davis, Chad, 199
dawgyg, 110, 112, 117–118
DDoS attacks. *See* distributed
denial-of-service attacks
DefCon, 11, 127, 136–141,
185, 201
CyberEthical Surfivor contest,
127–129, 140
Spot the Fed, 63
Web site, 209
Dell attack, 72, 75
denial-of-service attacks, 186, 200
dialed-number recorders, 80
diary entry, 182–184
Dick, Ron, 185–186
Diffie, Witt, 78
Digital Defense, 180–181
Dilley, Lloyd S., 200
Dispatchers, 112, 113
See also RaFa
distributed denial-of-service
attacks, 79, 192
Yahoo!, 67–70

DNRs. *See* dialed-number
recorders
Draper, John, 133, 190
Drunken Whores hacker club, 139

E

EHAP. *See* Ethical Hackers Against
Pedophilia
Erik Bloodaxe, 127
Ethical Hackers Against
Pedophilia, 26
ethics
CyberEthical Surfivor contest,
127–129, 140, 201
Ethical Hackers Against
Pedophilia, 26
hacker ethic, 11, 183
E-Trade attack, 70
Exodus, 89
Explotion, 90–91, 101–108
chronology, 198, 200, 201
current status, 108
early hacking exploits,
103–104
early life, 102–103
high school years, 104–106
online handle, 101–102
System Of A Down hack,
107–108

F

FBI
investigation of cyberwar
against NATO, 60–62
undercover operation to
penetrate hacking
community, 63–67
See also Computer Intrusion
Squad; National
Infrastructure Protection
Center
Ferris Bueller's Day Off (movie), 37
Finger o' Death, 131
Fleming, Rick, 166–170, 178–181

FonE_TonE, 110, 119–122
    chronology, 199
    See also World of Hell
friendships, 134

G

Genocide
    attacking pedophiles online,
        21, 188
    chronology, 196, 197
    Crack, 8–9
    current status, 26
    early life, 3–4
    early Unix hacking, 7–8
    first FBI contact, 15 19
    first major break-in, 15
    flame war, 22
    formation of hacker group,
        10–18
    high school hacking, 4–7
    manifesto, 20
    move to Oregon, 22
    Mr. Jerkins, 23–24
    nickname, 2–3
    second FBI contact, 23–25
    social engineering, 5
    See also Genocide2600
Genocide2600
    chronology, 197, 199
    current status, 26
    hacker ethic, 11
    manifesto, 20
    meetings, 12–13
    members of, 10–11
    national spread of, 22–23
    tools and scripts, 15
    Web site, 13–14, 22, 25, 208
Global Hell, 199
Goggans, Chris, 127
Gonzalez, Willie, 142–165, 192
    BBS surfing, 147
    chronology, 196, 197
    early life, 145–147
    high school years, 148–149

    introduction to hacking,
        142–144
    IRC Nuke Wars, 149–151
    Kryp, 153–155, 161–165
    school network hack, 151–153
    working as network security
        specialist, 156–161
Gosselin, Mark, 57–59, 76–77, 81,
    86, 89
Granick, Jennifer, 127–128
Gregory, Patrick W., 199
grimR, 120

H

H3C, Web site, 209
Hack in the Box, Web site, 210
hacker ethic, 184
    Genocide2600, 11
hacker friendships, 134
hacker groups
    Web sites, 208
    See also hacker group names
Hacker's Defense Foundation,
    Web site, 210
Hackers (movie), 197
Hackers.com: Next Evolution
    Hackers, Web site, 210
HackerWatch, Web site, 211
hacking
    justification of, 191
    as a public service, 190
Hacking For Satan, 93, 100
    Archdiocese of Baltimore
        alert, 96
    chronology, 201
    Web site defacement, *94*
    See also Pr0metheus
Hackology Network, Web site, 210
Hackweiser, 114
hactivism, 96
    See also activism
Harris, Eric, 1–2
headlines, chronology, 203–206
HV2K. *See* Sanford, Russell

I

Internet, growth of, 41
Internet Relay Chat Help, Web
  site, 211

K

Klebold, Dylan, 1–2
Knesek, Jill, 61–67, 70–71, 81, 86, 89
Kr0n, 122–124
  chronology, 198
  See also World of Hell
Kryp, 153–155, 161–165

L

L0pht, 78
  chronology, 198
L0phtCrack, 78, 156, 159
Ladopoulos, Eli, 35
LaVey, Anton S., 97, 99
Lederer, Richard, 93
Legion of Doom, 127, 129
  chronology, 196
Library, Web site, 211
LOD. *See* Legion of Doom
Lord Blackbane's Underground
  Lair, Web site, 207
Lulham, William G. "Greg", 200

M

Mafiaboy, 55–89, 188
  arrest, 86
  attacks on CNN, Datek, and
    E-Trade, 70
  chat log, 72–74
  chronology, 199, 200
  Dell attack, 72, 75
  e-mail accounts, 76
  informant on, 82
  and Kevin Mitnick, 88–89
  Operation Claymore, 59
  school experience, 84–85
  sentencing, 87–88

  wiretap on, 81–86
  Yahoo! DDoS attack, 67–70
  See also Outlawnet
Magee, Joe, 27–54
  2600 hacker group, 36
  BBS surfing, 34–35, 42–43
  chronology, 196
  current status, 53–54
  early hacking exploits, 41–42
  early life, 27–31
  first computer work, 33–34
  Masters of Deception, 35–36
  MLK Jr. Day telephone
    switching hack, 48–49
  ONIX, 35
  Stairwell 7, 36–37
  TRS-80, 34
  war dialing, 46–49
Malcom, 10–11, 12
manifestos
  false manifestos, 184
  Genocide2600, 20
Masters of Deception, 35–36, 129
  chronology, 196
Mitnick, Kevin, 191–192
  chronology, 195, 197, 199
  and Mafiaboy, 88–89
  Web site, 207
Miville-Deschenes, Louis, 88
Mixter, 76
MOD. *See* Masters of Deception
Moore, Anna, 192
  2600 hacker group, 135–136
  chronology, 197, 199, 201
  current status, 141
  CyberEthical Surfivor contest,
    127–129, 140
  DefCon, 136–141
  early life, 129–131
  Internet handle, 131–132, 133
  learning Linux and C,
    133–134
  Nuke Wars, 131–133

parents of, 129–132
phone phreaking, 133
Moore, H.D., 166–181
chronology, 196, 197–198, 199
Computer Sciences
Corporation, 166–170,
178–180
Digital Defense, 180–181
discovery of AOL
vulnerabilities, 173–174
early hacking exploits, 172–174
early life, 170–172
first major hacking program,
175
high school, 175, 177
NLog, 168, 176, 177
Nuke Wars, 174
phpDistributedPortScanner,
179
Shadow project, 168–170, 177
war dialing, 173
Web site, 207
Moran, Dennis, 199
MostHateD. See Gregory, Patrick W.
Mr. Jerkins, 23–24
Mshadow, 72
Mudge, 78–79

N

NASA hacks, 98
cost of, 182
by Raymond Torricelli, 189
National Infrastructure Protection
Center, 185
DDoS warnings, 80
Web site, 211
National Petroleum Technology
Office hack, 118
NATO
bombing of Chinese embassy,
59–60
cyberwar against, 60
Neal, Charles, 61, 64, 66, 70–72, 89

Netcat, 83
NIPC. See National Infrastructure
Protection Center
NLog, 168, 176, 177, 198
Nmap, 169–170, 176
Noid, 27–54
alternate online existence,
38–39
Apocalypse, 43–46, 49–50
change in the hacker scene,
49–52
chronology, 196
current status, 54
early problems with the
law, 37
end of hacking, 52–53
first computer, 37–38
introduction to electronics,
31–33
IRC, 52–53
Revenge, 43–46, 49–51
screen name, 38
software cracking, 39–40, 43–46
Noonan, Tom, 78
Nuke Wars, 131–133, 149–151,
174, 197
NukeNabber, 132

O

Oklahoma. See Moore, Anna
ONIX, 35
Operation Claymore, 59
first hacks, 64–67
See also Mafiaboy
Operation Sundevil, 129
Outlawnet, 55–57, 77, 87
See also Mafiaboy

P

Patriot Act, 187
pedophilia
attacking online, 21
Ethical Hackers Against
Pedophilia, 26

pen registers. *See* dialed-number
  recorders
Pentagon hacks, cost of, 182
Perl, 91
Pethia, Rich, 78
ph33r hax0r, 117
Ph33r-the-B33r, 112, 114
Phiber Optik. *See* Abene, Mark
phone phreaking, 5, 133, 190
phpDistributedPortScanner, 179
Phrack Magazine, Web site, 212
Ping o' Doom, 131, 132
ping-of-death attack, 60
PoizonB0x, 98–100
    chronology, 201
pornography, child, 26, 188
Pr0metheus, 90–101, 188
    chronology, 201
    current status, 101
    Defense Information Systems
      Agency hacks, 98
    early hacking exploits, 96–97
    Los Alamos National
      Laboratory hacks, 98–99
    message, 92–93
    PoizonB0x, 98–100
    Satanism, 96, 97–98
    Sigil of Baphomet, 94–95
    Washington State Court
      hacks, 98–99
    Web site defacement, *94*
    See also Hacking For Satan
Princess Zelda. *See* Moore, Anna
Prophet, 183

R

r00t, 110
r00t-access, 120
RaFa, 110, 112, 113, 116–119,
  124–126
    chronology, 200
    hacker handle, 116–117

holiday greeting, *125*
    See also Dispatchers; World
      of Hell
Randomizer, 76
Reb. *See* Harris, Eric
Renner, Jon, 55–57
Report a Cybercrime, Web site, 212
Revenge, 43–46, 49–51
Rolex. *See* Torricelli, Raymond
Romanowski, Yan, 87
Royal Canadian Mounted Police
  (RCMP), 57

S

San Jose State University hacks, by
  Raymond Torricelli, 189
Sanford, Russell, 200
Satanism. *See* Hacking For Satan;
  Pr0metheus
Schanot, Christopher, 197
Schwartau, Winn, 127, 129
Scorpion, The. *See* Stira, Paul
script kiddies, 67, 185
    See also teenage hackers
Secondary Heuristic Analysis
  Systems for Defensive Online
  Warfare. *See* Shadow project
September 11 terrorist attack, 2,
  109, 110, 186
Shadow Hawk. *See* Zinn, Herbert
Shadow project, 168–170, 177
Skeleton Crew, The, 184
Social Base of the Hacker, The, 20, 197
social engineering
    contest, 139
    Genocide, 5
software piracy, 43–46
Sonic Hacking Community, Web
  site, 210
Spirit. *See* Gonzalez, Willie
Sprint Canada, 57, 58
Stacheldraht, 75

Staples Online hack, 121
Star Road. *See* Moore, Anna
Starla Pureheart. *See* Moore, Anna
Stira, Paul, 35
Swallow, Bill, 60–67, 89
swinger, 72

T

T3, 72
tar ball, 23
Teen Programmers Unite,
    Web site, 212
teenage hackers, 187–193
    parents of, 188–189
    public schools, 189
telephone switching, 47–49
Tenebaum, Ehud, 198
terrorism. *See* cyberterrorism
Texas State Lottery hack, 118
Texts Library, Web site, 211
TFN. *See* Tribal Flood Network
The Satanic Bible, 97
Thieme, Richard, 140
ToneLoc, 173
Torricelli, Raymond, 189–190
    chronology, 200
Tribal Flood Network, 75

U

U.S.A. Patriot Act, 187

V

VCRs, development of, 31
VoDKa. *See* Klebold, Dylan
Vranesevich, John, Web site, 207

W

W00w00 Security Development,
    Web site, 212

Walker, John, 187
war dialing, 197
    Magee, Joe, 46–49
    Moore, H.D., 173
War Games (movie), 37, 47, 195
Web sites
    hacker groups, 208
    hacking information,
        209–211
    individual hackers,
        207–208
    security information,
        211–212
Well, The, 34–35
white-hat hacker community, 15
WinNuke, 149–150, 151
Wiser, Leslie, 185
WiZDom, 10, 12
World of Hell, 113, 114–116
    chronology, 198, 199, 200
    holiday greeting, *125*
    mission, 118
    trademark defacement, *109*
    See also Cowhead2000;
        dawgyg, FonE_TonE; Kr0n;
        RaFa
World Trade Center attack, 2, 186
Wozniak, Steve, 33, 133, 183, 190
    Web site, 208

Y

Yahoo! DDoS attack, 67–70

Z

Zatko, Peiter "Mudge", 78–79
Zinn, Herbert, 196
zombie machines, 70, 113, 186
Zyklon. *See* Burns, Eric

INTERNATIONAL CONTACT INFORMATION

AUSTRALIA
McGraw-Hill Book Company Australia Pty. Ltd.
TEL +61-2-9417-9899
FAX +61-2-9417-5687
http://www.mcgraw-hill.com.au
books-it_sydney@mcgraw-hill.com

CANADA
McGraw-Hill Ryerson Ltd.
TEL +905-430-5000
FAX +905-430-5020
http://www.mcgrawhill.ca

GREECE, MIDDLE EAST,
NORTHERN AFRICA
McGraw-Hill Hellas
TEL +30-1-656-0990-3-4
FAX +30-1-654-5525

MEXICO (Also serving Latin America)
McGraw-Hill Interamericana Editores S.A. de C.V.
TEL +525-117-1583
FAX +525-117-1589
http://www.mcgraw-hill.com.mx
fernando_castellanos@mcgraw-hill.com

SINGAPORE (Serving Asia)
McGraw-Hill Book Company
TEL +65-863-1580
FAX +65-862-3354
http://www.mcgraw-hill.com.sg
mghasia@mcgraw-hill.com

SOUTH AFRICA
McGraw-Hill South Africa
TEL +27-11-622-7512
FAX +27-11-622-9045
robyn_swanepoel@mcgraw-hill.com

UNITED KINGDOM & EUROPE
(Excluding Southern Europe)
McGraw-Hill Education Europe
TEL +44-1-628-502500
FAX +44-1-628-770224
http://www.mcgraw-hill.co.uk
computing_neurope@mcgraw-hill.com

ALL OTHER INQUIRIES Contact:
Osborne/McGraw-Hill
TEL +1-510-549-6600
FAX +1-510-883-7600
http://www.osborne.com
omg_international@mcgraw-hill.com